# Arctic and Alpine Biomes

# GREENWOOD GUIDES TO
# BIOMES OF THE WORLD

**Introduction to Biomes**
Susan L. Woodward

**Tropical Forest Biomes**
Barbara A. Holzman

**Temperate Forest Biomes**
Bernd H. Kuennecke

**Grassland Biomes**
Susan L. Woodward

**Desert Biomes**
Joyce A. Quinn

**Arctic and Alpine Biomes**
Joyce A. Quinn

**Freshwater Aquatic Biomes**
Richard A. Roth

**Marine Biomes**
Susan L. Woodward

# Arctic and
# ALPINE BIOMES

Joyce A. Quinn

**Greenwood Guides to Biomes of the World**

Susan L. Woodward, General Editor

GREENWOOD PRESS

Westport, Connecticut • London

**Library of Congress Cataloging-in-Publication Data**

Quinn, Joyce Ann.
  Arctic and alpine biomes / Joyce A. Quinn.
    p. cm.
  Includes bibliographical references and index.
    ISBN 978–0–313–33840–3 (set : alk. paper) — ISBN 978–
0–313–34017–8 (vol. : alk. paper)
    1. Tundra ecology. 2. Tundra ecology—Arctic regions. 3.
Mountain ecology—Alps Region. 4. Alpine regions. I. Title.
  QH84.1.Q56 2008
  577.5′86—dc22        2008027510

British Library Cataloguing in Publication Data is available.

Library of Congress Catalog Card Number: 2008027510
ISBN: 978–0–313–34017–8 (vol.)
        978–0–313–33840–3 (set)

First published in 2008

Greenwood Press, 88 Post Road West, Westport, CT 06881
An imprint of Greenwood Publishing Group, Inc.
www.greenwood.com

Printed in the United States of America

∞

The paper used in this book complies with the
Permanent Paper Standard issued by the National
Information Standards Organization (Z39.48–1984).

10  9  8  7  6  5  4  3  2  1

# Contents

# Preface

In the course of traveling throughout the United States and the world, I have been fortunate to experience arctic and alpine environments firsthand. I have visited Lapland and the Land of the Midnight Sun, climbed mountains in the Sierra Nevada and the Pyrenees, and hiked over many high passes in the mountains of North America, Europe, and Asia. Walking through a flower-filled meadow and hearing the whistle of a marmot from a rock pile is a treat for the senses.

This book can be only an introduction to a vast topic that would take volumes to thoroughly explore. It covers a small part of the variety in the natural world, much of which is at risk because of human activities, whether purposeful or inadvertent. A little knowledge of a subject arouses curiosity—a desire to learn more. And as one learns more, the subject takes on greater importance. Conservation issues become more meaningful when the public is aware of the interactions among elements in the natural world. Hopefully, the information in these pages will instill an appreciation for arctic and alpine environments and their natural inhabitants. Instead of seeing individual plants in a botanic garden or animals in captivity, readers will have a feeling for the value of each in its natural place in the world.

Arctic and alpine biomes are the cold regions beyond the limit of tree growth, either because of latitude or elevation. The first chapter explains elements, such as temperature, precipitation, and forms of life, common to the Arctic, Antarctic, and alpine regions. It also mentions differences, which are elaborated on in subsequent chapters. Selected geographic regions are described in chapters dealing with Arctic and Antarctic, mid-latitude alpine environments, and tropical alpine regions.

Textual material is illustrated with numerous maps, diagrams, photographs, and line drawings. The intended audience is not only advanced middle school and high school students, but also university undergraduates and anyone else who is interested in the natural environment of high mountains.

I would like to thank Kevin Downing of Greenwood Press for his insights and constant support in the completion of this project. Jeff Dixon did a masterful job of deciphering my sketches to produce meaningful illustrations. Bernd Kuennecke of Radford University's Geography Department prepared the maps of the worldwide distribution of arctic and alpine regions. Several people generously provided pictures for the book, while others read drafts and offered suggestions. I am deeply indebted to all individuals who enabled me to finalize this volume. Any errors, however, are solely my own.

# How to Use This Book

The book is arranged with a general introduction to arctic and alpine biomes and a chapter each on the Arctic and Antarctic Tundra Biome, Mid-latitude Alpine Tundra Biome, and the Tropical Alpine Environment Biome. The introduction describes unifying characteristics, such as the physical environment of the biome and plant and animal adaptations. Subsequent chapters begin with a general overview at a global scale and descriptions of features specific to that biome. Regional descriptions are organized by the continents on which they appear. Each chapter and each regional description can more or less stand on its own, but the reader will find it instructive to investigate the introductory chapter and the introductory sections in the later chapters. More in-depth coverage of topics perhaps not so thoroughly developed in the regional discussions usually appears in the introductions.

The use of Latin or scientific names for species has been kept to a minimum in the text. However, the scientific name of each plant or animal for which a common name is given in a chapter appears in an appendix to that chapter. A glossary at the end of the book gives definitions of selected terms used throughout the volume. The bibliography lists the works consulted by the author and is arranged by biome and the regional expressions of that biome.

All biomes overlap to some degree with others, so you may wish to refer to other books among Greenwood Guides to Biomes of the World. The volume entitled *Introduction to Biomes* presents simplified descriptions of all the major biomes. It also discusses the major concepts that inform scientists in their study and understanding of biomes and describes and explains, at a global scale, the environmental factors and processes that serve to differentiate the world's biomes.

# The Use of Scientific Names

Good reasons exist for knowing the scientific or Latin names of organisms, even if at first they seem strange and cumbersome. Scientific names are agreed on by international committees and, with few exceptions, are used throughout the world. So everyone knows exactly which species or group of species everyone else is talking about. This is not true for common names, which vary from place to place and language to language. Another problem with common names is that in many instances European colonists saw resemblances between new species they encountered in the Americas or elsewhere and those familiar to them at home. So they gave the foreign plant or animal the same name as the Old World species. The common American Robin is a "robin" because it has a red breast like the English or European Robin and not because the two are closely related. In fact, if one checks the scientific names, one finds that the American Robin is *Turdus migratorius* and the English Robin is *Erithacus rubecula*. And they have not merely been put into different genera (*Turdus* versus *Erithacus*) by taxonomists, but into different families. The American Robin is a thrush (family Turdidae) and the English Robin is an Old World flycatcher (family Muscicapidae). Sometimes that matters. Comparing the two birds is really comparing apples to oranges. They are different creatures, a fact masked by their common names.

Scientific names can be secret treasures when it comes to unraveling the puzzles of species distributions. The more different two species are in their taxonomic relationships the farther apart in time they are from a common ancestor. So two species placed in the same genus are somewhat like two brothers having the same father—they are closely related and of the same generation. Two genera in the same family might be thought of as two cousins—they have the same grandfather, but different fathers. Their common ancestral roots are separated farther by time. The important thing in the study of biomes is that distance measured by time often means distance measured by separation in space as well. It is widely held that new species come about when a population becomes isolated in one way or another from the rest of its kind and adapts to a different environment. The scientific classification into genera, families, orders, and so forth reflects how long ago a population went its separate way in an evolutionary sense and usually points to some past environmental changes that created barriers to the exchange of genes among all members of a species. It hints at the movements of species and both ancient and recent connections or barriers. So if you find two species in the same genus or two genera in the same family that occur on different continents today, this tells you that their "fathers" or "grandfathers" not so long ago lived in close contact, either because the continents were connected by suitable habitat or because some members of the ancestral group were able to overcome a barrier and settle in a new location. The greater the degree of taxonomic separation (for example, different families existing in different geographic areas) the longer the time back to a common ancestor and the longer ago the physical separation of the species. Evolutionary history and Earth history are hidden in a name. Thus, taxonomic classification can be important.

Most readers, of course, won't want or need to consider the deep past. So, as much as possible, Latin names for species do not appear in the text. Only when a common English language name is not available, as often is true for plants and animals from other parts of the world, is the scientific name provided. The names of families and, sometimes, orders appear because they are such strong indicators of long isolation and separate evolution. Scientific names do appear in chapter appendixes. Anyone looking for more information on a particular type of organism is cautioned to use the Latin name in your literature or Internet search to ensure that you are dealing with the correct plant or animal. Anyone comparing the plants and animals of two different biomes or of two different regional expressions of the same biome should likewise consult the list of scientific names to be sure a "robin" in one place is the same as a "robin" in another.

# 1

# Introduction to Arctic and Alpine Biomes

Arctic and alpine biomes, also called tundra, are treeless regions characterized by several factors that limit plant growth, including low temperatures, shallow nutrient-poor soils, low precipitation, and drying winds. Incomplete plant cover gives the impression of a desert, but it is not always dry. Low temperatures limit evaporation, and there may be moisture or ice deeper in soil and rock. Although it does not refer to the poles but to high latitudes near and above the Arctic and Antarctic Circles, "polar" tundra may be a better term for treeless areas in high latitudes because it includes the Antarctic as well as the Arctic. Unless otherwise specified, arctic is used here to refer to both polar regions. Alpine refers to treeless areas at high elevations in mountains at any latitude.

It is often stated that alpine climate and vegetation on mountain tops is the same as in the arctic tundra. Mean annual temperature and duration of snow cover may be similar, but differences outnumber similarities (see Table 1.1). Both arctic and mid-latitude alpine tundra have mean annual temperatures below freezing. Because of their mid- to high-latitude positions, both have extreme differences between summer and winter temperatures. Winters can be very cold with temperatures never rising above freezing, while summers are consistently cool. Summers average less than 50° F (10° C) even though air temperature during the day may reach 60°–70° F (15.5°–21° C). The growing season, roughly the time between the last substantial frost in spring and the first in fall is 6–10 weeks, but freezing temperatures are possible even during the brief summer. Soils are shallow and nutrient poor, with variations in rockiness and drainage. Both types of tundra have plant life dominated by dwarf shrubs, mosses, lichens, sedges, and perennial forbs and both

**Table 1.1 Environmental Comparison between Arctic Tundra and Alpine, Both Mid-Latitude and Tropical**

| Characteristic | Arctic Tundra | Mid-Latitude Alpine | Tropical Alpine | Antarctic Tundra |
|---|---|---|---|---|
| Treeline | Sea level to 330 ft (100 m) | Average for 40°–50° lat N – 5,000–9,850 ft (1,500–3,000 m) S – 3,300–6,550 ft (1,000–2000 m) | 9,850–13,000 ft (3,000–4,000 m) | No trees |
| Permafrost | Continuous and discontinuous | Discontinuous | None | Rare |
| Length and daily mean temperature of growing season | Short, with low temperature | Short, with low temperature | Year-round, with low temperature | Short, with low temperature |
| Diurnal temperature variation | Little change because there is no day or night | Extreme depending on latitude and light regime | Extreme | Little change because there is no day or night |
| Seasonal temperature change | Extreme | Extreme | Little | Extreme |
| Topography | Level to rolling | Bare rock, cliffs, deep glacial valleys | Bare rock, cliffs, deep glacial valleys | Level to rolling, some mountains, valleys, and nunataks |
| Soils | Shallow and nutrient poor or boggy | Shallow and nutrient poor | Shallow and nutrient poor | Shallow and nutrient poor |
| Soil drainage | Impeded by permafrost | Depends on topography | Depends on topography | Well drained |
| Light regime (day length) | 24 hours of daylight or darkness depending on season | Hours of daylight and darkness vary according to latitude | 12 hours of daylight and darkness all year | 24 hours of daylight or darkness depending on season |
| Sunlight intensity and ultraviolet light | Low, especially when cloudy | Intense except when cloudy | Intense except when cloudy | Low, especially when cloudy |
| Atmospheric content | Normal | Decreasing $O_2$ and $CO_2$ with elevation | Decreasing $O_2$ and $CO_2$ with elevation | Normal |

| | | | | |
|---|---|---|---|---|
| Pressure | Normal | Decreasing with elevation | Decreasing with elevation | Normal |
| Microclimate in sun vs. shade | Minor | Extreme depending on elevation | Extreme depending on elevation | Minor |
| Wind | Calm to strong depending on location of Arctic Front | Depends on topography and latitude | Calm | Calm to strong depending on location of Antarctic Front |
| Precipitation | Low | Increasing with elevation | Increasing with elevation to a cloud level, then decreasing | Low |
| Snow cover | Thin | Varies, thin to deep | Varies, thin to deep | Thin to none |

**Sun or Shade?**

Most of the energy coming from the sun, short wavelength, freely passes through Earth's atmosphere to the ground where it is absorbed. The Earth then reradiates that energy in longer wavelengths, called infrared. The atmosphere does not allow longer wavelengths to easily escape back to space, instead absorbing the energy and keeping the air warm. At high elevations where the air is thinner, even more of the shortwave energy is absorbed at ground level, raising surface temperatures in sunny locations. When you stand in sunlight on a high mountain, you feel warmth because your body is absorbing solar radiation. Because fewer atmospheric particles are in the thin air, however, more infrared energy radiates through the air and back to space, leaving the surface and air cold. Step into the shade and you feel the cooler air temperature. Alpine vegetation experiences these temperature extremes daily. Fog or clouds moderate temperature by blocking both incoming shortwaves and outgoing longwaves.

have patterns of wet meadows, dry heath or shrubs, and rocky habitats according to microclimate and substrate characteristics. Alpine vegetation patterns are on a smaller scale, with less extensive wet tundra because of less permafrost (permanently frozen soil) and better drainage. Vegetation belts in the Arctic and Northern Hemisphere mid-latitude alpine tundra also have similar floras. Zones of vegetation successively change from forests to stunted trees (krummholz) to tall shrubs to low-stature tundra plants, often with the same genera or species. Alpine vegetation in the Tropics and in the Southern Hemisphere, however, is quite different. The biggest similarity among all arctic and alpine tundra is the lack of trees, which together with dominance of particular growthforms and generally cool temperatures categorize these environments into the same biome.

Differences between arctic and alpine tundra are also numerous. Light regime and receipt of solar radiation are vastly different. Arctic regions experience up to six months of daylight or darkness, while alpine regions have seasonality in daylength and sun angle according to latitude. Seasonal temperature variations in polar regions are much greater than diurnal variations. The opposite is true for alpine areas in the Tropics; temperatures change more from day to night than they do from winter to summer. Because no standard daylight and night exist for a 24-hour period in the Arctic, temperature remains the same, which is especially important during the growing season in the summer with perpetual daylight. At high elevations, incoming solar radiation is intense and temperatures are high during the day, while at night, infrared energy is more easily lost to space, causing temperatures to drop. Climate characteristics of alpine locations in the Tropics are distinctive because of a lack of seasonality. Tropical alpine environment temperature regimes are sometimes described as summer every day and winter every night. Alpine areas have greater variety in local climates related to slope steepness and aspect. Steep slopes may intercept low-angle solar radiation at a more direct angle, thereby absorbing more energy than a flat area would. Depending on latitude and position north or south of the Equator, north-facing or south-facing slopes may be always sunny, always shady, or variable according to season. Some, but not all, alpine areas have strong winds, while wind in the Arctic is usually negligible. Little snow falls in

Arctic regions, while some alpine areas receive several feet each winter. Permafrost is often lacking in alpine environments, as are large-scale wet habitats caused by poor drainage. Long Arctic nights may put constraints on animal life, while thin atmosphere at high elevations affects both plants and animals.

The term tundra is not easily defined, and it becomes even more problematic at the Equator and in the Southern Hemisphere. The Russian word *tundra,* meaning land of no trees, may be derived from either the Finnish *tunturi* meaning barren lands or from the Lapp *tundar* meaning marshy plain. Strictly speaking, the term should be used to refer to treeless landscapes in the Arctic or Antarctic as opposed to mountain tops

The term alpine stems from the European Alps and is frequently used to describe entire mountain areas. In a more restricted sense, it refers to areas on mountains above the level where trees grow. The term alpine tundra is also used for treeless areas at high elevation. Many scientists limit the term alpine to refer to mountains with steep slopes, excluding high-elevation flat plateaus such as Tibet at 16,400 ft (5,000 m). Because alpine areas are found in all parts of the world—polar, tropical, oceanic islands, continental areas, wet, and dry regions—climate varies widely. Alpine tundra can be subdivided into two major regions: Arctic-alpine tundra associated with mid-latitude mountains in the Northern Hemisphere and tropical alpine environments on mountains in the Tropics or Subtropics, primarily in South America and Africa. Smaller alpine areas on high mountains in the Southern Hemisphere are a minor third subdivision. Because of north-south migration corridors in the Northern Hemisphere, alpine tundra in those regions bears many similarities to the Arctic, including many of the same plant species. In contrast, tropical mountains and Southern Hemisphere alpine regions have unique floras because of their isolation.

## Location

Arctic landscapes are circumpolar, surrounding the North Pole on both North American and Eurasian continents and on Arctic islands. Alpine tundra has a global distribution and is found at all latitudes, but it can be fragmented and confined to isolated mountain peaks. Northern Hemisphere alpine environments merge with circumpolar tundra where mountains extend to high latitudes, but they are isolated on mountain ranges as far south as Nepal and southwestern China. Except for areas used for mountain pasture, remoteness and inaccessibility of alpine areas has helped preserve native and natural vegetation. High mountains are located in many floristic provinces, which is one factor that contributes to plant diversity; but diversity is also enhanced by geographic isolation, climatic change, mountain building, glaciation, and microhabitats. A small-scale map showing tundra may be misleading because delimitations are frequently broad ecotones rather than abrupt boundaries, and mountain top areas are often too small to be depicted accurately.

### Definition and Boundaries

Terminology used by various scientists to refer to zones of either arctic or alpine tundra is not consistent. Terms such as hemi-arctic tundra, southern tundra, mid-alpine, or low-alpine are imprecise. No precise meaning exists for even the simpler terms of arctic, subarctic, alpine, and subalpine, and not all authors or scientists use the same definitions. Arctic stems from the Greek word for bear, *Arktos,* and is now used to refer to the high-latitude lands under the North Star in the Ursa Major and Ursa Minor constellations. Alpine comes from the Latin *Alpes* for snow-covered or white mountains north of Italy, the Alps. The "sub" preface, meaning "not quite" infers less of the conditions to make it full arctic or full alpine tundra. Subarctic and subalpine are often slightly equatorward in latitude or just below tundra in elevation. Antarctic and subantarctic refer to the Southern Hemisphere equivalent of arctic and subarctic.

In terms of latitude, arctic refers strictly to the area between the Arctic Circle and the North Pole, but in terms of biomes, it means the treeless zone, which may extend south of the Arctic Circle. Tundra may or may not be underlain by permafrost. Boundaries of permafrost sometimes correlate with treeline, but some permafrost may be a factor of past, not present, climates.

Alpine areas are most often defined by treeline because climate data are sparse in mountains. The use of treeline as a boundary is generally satisfactory in the Northern Hemisphere but this criterion cannot be applied worldwide. Tall rosette plants in tropical mountains make definition of trees and treeline difficult. Because the western slopes of the Andes rise from desert rather than forest, no treeline can occur. The 50° F (10° C) summer isotherm is an often-used criterion. It sometimes coincides with tundra vegetation because trees cannot grow with cooler summer temperatures. Subarctic and subalpine refer to the krummholz zone, where tree species are dwarfed by climatic conditions, but these areas also include sheltered groups of full-size trees in the Arctic.

High arctic and polar desert refer to bare rocky areas supporting only lichens, with small pockets of bog or flowering plants in favorable or sheltered habitats. Equivalent in high mountains is the nival or aeolian zone above the limit of vascular plants. Although the terms are often used interchangeably, nival refers to snow-fields and exposed rock, while aeolian refers to high-elevation environments to which wind transports insects and nutrients. The aeolian zone is most extensive in the Andes and Himalayas because of their higher elevations.

Indicator plants may be used to delimit tundra. Typical arctic-alpine species such as moss campion, alpine sorrel, spike trisetum, and purple saxifrage, which can withstand cool summers and winter winds, are good indicators of Northern Hemisphere tundra (both arctic and alpine) but not of tropical or Southern Hemisphere alpine zones.

## Treeline

The term treeline is indistinct, especially in places where human history of forest destruction blurs the boundary between natural and artificial and results in an increase in treeless extent, as in Eurasia. Treeline refers to the transition zone between forest and the arctic or alpine environment where trees, defined as single upright stems at least 20 ft (3 m) tall, cease to grow. Some of the same tree species may continue to occur above treeline as dwarfed or prostrate shrubs. The term timberline, referring to sizeable logs useable for timber, is not synonymous with treeline, but the two are often used interchangeably. Treeline varies with latitude, generally occurring at sea level to 330 ft (100 m) in the Arctic or subarctic to 9,850–13,000 ft (3,000–4,000 m) in the Tropics. In the mid-latitudes, treeline is higher in the Northern Hemisphere, 5,000–9,850 ft (1,500–3,000 m) at 40°–50° N compared with 3,300–6,550 ft (1,000–2,000 m) at 40°–50° S (see Figure 1.1).

Several factors are responsible for variation in treeline, especially in the Northern Hemisphere mid-latitudes. Slope aspect, coastal versus inland location, and precipitation add climate variables other than those caused by latitude. Treeline is highest in continental areas because of more accumulated summer heat. Snowline shows a similar pattern, again with more variation in the Northern Hemisphere.

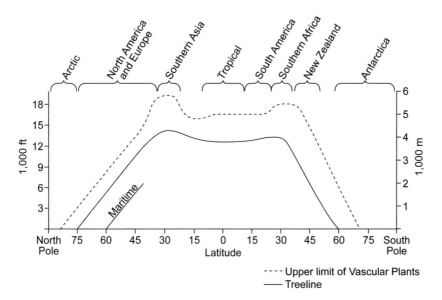

**Figure 1.1** The elevation of treeline generally increases from the Arctic and Antarctic toward the Equator. However, it peaks in the low mid-latitudes and declines slightly in the Tropics. Treeline is higher in the Northern Hemisphere because of larger landmasses that retain more heat. The highest limit of vascular plants follows a similar pattern. *(Illustration by Jeff Dixon. Adapted from Price 1981.)*

Many factors such as temperature, elevation, daylength, and growing season are interrelated, making it impossible to conclude that one factor is responsible for treeline. Treeline is controlled by wind, snowdrifts, rocks, and soils, but trees also affect the local climate by shading the ground, trapping snow, providing wind protection, intercepting ultraviolet light, and moderating diurnal temperature changes. Factors significant to treeline in the mid-latitudes may be irrelevant in the Tropics. Although only mean values are often available, extremes or duration of climate factors are usually more significant. Past climates may determine the position of current treeline because trees live for hundreds of years, but they may not reproduce under current conditions. It is frequently stated that a mean temperature of the warmest month of 50° F (10° C) is the limit for tree growth. While this figure may be accurate for temperate zones, a mean temperature of 43° F (6° C) for the growing season, a period ranging from 2.5–12 months, may be a better worldwide measure. During the Pleistocene, temperate zone mountains were deforested and the alpine zone extended to the lowlands, providing migration routes for arctic and alpine plants. In the Tropics, treeline was 3,300–5,250 ft (1,000–1,600 m) lower.

With one possible exception, no one factor can account for treelines worldwide. Why trees do not grow above a certain elevation involves possible causes such as depth and extent of snow cover, which may shorten the growing season too much. Deep, long-lasting snow may also harbor molds and fungi, which might inhibit plant growth. Wind subjects plants to stress, abrasion, blowing ice, and desiccation. Wind can also blow seeds away from suitable germination sites. Biotic activities, such as animals eating bark, seeds, or seedlings, or insect infestations are other factors that can be responsible for the development of treeline. Human activities such as repeated burning or cutting trees for timber or firewood may have devastated natural treeline in some areas. Variations in solar radiation intensity or cloud cover are yet other possibilities. A common explanation is damage from repeated freezing; however, treelines form in maritime mountain climates where frost is rare. Similarly, mechanical breakage by ice and wind is a potential factor, but treelines occur in regions that have little wind and ice-scouring. A short growing season that limits reproduction may contribute to treeline in mid-latitude mountains, but the concept cannot be applied to tropical mountains that have no seasonality.

Trees need time for soft new tissue or shoots to complete growth and physiologically winter harden, so they become able to withstand low temperatures and desiccation. Trees with a high canopy prevent solar energy from penetrating down to the soil. At high elevations and cooler temperatures, the root zone does not become warm enough to allow time for winter hardening and growth is limited. Small-stature plants in the alpine zone actually have higher temperatures beneath their canopies, which are lower to the ground, than do taller trees. One response is for individual trees to be more widely spaced at treeline, which allows the root zone to become warmer.

Temperature, specifically summer temperatures in the mid-latitudes, usually determines treeline, but aridity such as in the Andean puna or in central Tibet is also a major factor. Treeline is higher than 14,750 ft (4,500 m) in Tibet and almost 16,400 ft (5,000 m) in the South American puna. Snowlines in those regions are higher than 19,700 ft (6,000 m), also because of aridity.

Treeline species in the Northern Hemisphere are mostly conifers, dominated by pines, spruce, and fir. Warmer areas of Eurasia and western North America also have larch, the only deciduous conifer. Birch is more common in cool, moist climates such as Fennoscandia (the peninsula that includes Norway, Sweden, and Finland); and alder and beech are prominent in central Europe and the Caucasus. Rhododendrons occur at treeline in both Europe and the Himalayas. Tropical and Southern Hemisphere treeline species are different because of their isolation from each other. The tropical high Andes have *Polylepis*, whereas a deciduous southern beech dominates in the southern Andes. New Zealand's treeline is dominated by *Podocarpus* and southern beech. Treeline in tropical Africa mountains is represented by ericaceous shrubs of tree heather and *Philippia* species, although the zonation pattern is complicated by tree-like plants that occur in the alpine zone. The highest tree on the dry Drakensburg Plateau in southern Africa is a *Protea*.

## Physical Environment

### Role of Glaciation

***Continental glaciation.*** Continental glaciers, or ice sheets, influenced the landscape of not only the arctic and antarctic tundra but also much of northern North America and Eurasia as far south as the middle latitudes in the United States, Europe, and Russia. The western part of North America and the eastern part of Asia, however, were too dry for glaciers to advance that far south. Several times during the Pleistocene, the climate in the Northern Hemisphere cooled sufficiently that winter snowfall failed to melt completely the following summer. Over thousands of years, the snow accumulated and compressed into ice more than a mile thick. The sheer weight depressed the center of the ice mass causing the edges to move away in all directions. Ice must move over land to be called a glacier, while the same ice mass moving into the ocean is called an ice shelf or ice floe as it floats on water. It is important to remember that in a glacier, ice is always moving away from the center; it never moves backward, toward the area of snow accumulation. If the rate at which snow and ice accumulates exceeds the rate of loss (ablation by melting, calving, and sublimation) along the edges of the ice, the ice front advances and moves farther away from the center.

**How Can Ice Move?**

Solid ice in a continental glacier moves very slowly across the landscape similar to how semisoft butter or bread dough would move across a table if pressure were applied to the center of the mass. Ice moves downslope in a valley glacier in response to gravitational pull. The manner of ice movement is complex but includes sliding over a thin layer of water melted from friction, like a skier on snow, and by ice sheer, similar to how each card in a deck of cards will slide over the others when pressure is applied to the side of the deck.

If the rate of accumulation equals the rate of ablation, the ice front remains stationary even though the ice itself is still moving forward. When the rate of accumulation is less than the rate of ablation, the ice front retreats, or melts back, toward the center of its origin. These movements occur today in the glaciers covering Antarctica and Greenland.

Specific movements of the ice did different things to the underlying landscape (see Plate I). As the ice advanced, particularly in Canada near its origins in North America, it scraped and gouged the land, scouring rocks bare of soil. At the edges of the ice and during times of glacial retreat, the debris of rock, sand, and clay (glacial drift) that was carried in the ice was deposited as the ice melted. Material deposited directly by the ice was not sorted; it became a jumble of different sizes called boulder clay. Hills and rolling landscapes of boulder clay are called moraines, with several different types depending on specific ice movement. Rivers and streams carrying meltwater away from the glacier sorted sediments according to size, leaving behind flat sandy or gravelly landscapes generally called outwash plains. Smaller areas of sorted sediment within and between moraines were sometimes caused by temporary lakes. Moraines composed of unsorted boulder clay have variations in drainage and often contain lakes and bogs. Lake beds may be underlain by poorly drained clay deposits, while sandy and gravelly outwash is usually well-drained habitat.

Except on highlands above the ice, called nunataks, continental ice sheets obliterated all living things on the landscape, forcing arctic plants and animals to move farther south as the proximity of ice cooled the adjacent climate over a time span of thousands of years. The last continental glaciers melted away in the Arctic less than 10,000 years ago, leaving little geologic time for plants to recolonize some areas.

***Alpine glaciation.*** The major difference between continental and alpine glaciers is that alpine glaciers do not cover the entire landscape but are confined to valleys (see Plate II). Snow and ice accumulates in the upper reaches of steep river valleys with a V-shape profile. As the ice moves downslope, it carves the sides of the valley as well as the bottom, resulting in a U-shaped profile with steep, sometimes vertical sides. As glaciers enlarge and erode more of their valleys, the original rounded landscape of the uplands between the valleys is eaten way by frost action, eventually becoming steep, rocky, serrated ridges called aretes. A rich terminology exists to describe various alpine glacial landforms. The amphitheater-shaped depression at the headwaters of the valley where the glacier began is a cirque, which may contain a lake called a tarn. Moraines develop along the valley walls and at the furthest extent of ice advance, although meltwater from the ice usually destroys and redistributes the debris in the center of the valley. The rounded, deep valley floor may partially fill with outwash sediments and become quite flat. Uplands and aretes are usually dry and windy, while valley floors may have wet habitats such as meadows and bogs.

When arctic plants followed continental glacial retreat back north, they also retreated upward into the mountains and found refuge during warm interglacial periods in the cooler climate at higher elevations. Isolation in different mountain ranges and different habitats in the alpine glacial landscape afforded opportunities for evolution and speciation.

## Snow

Snow is possible at any time of the year at high latitudes and elevations, even in tropical alpine environments. Because some snow may persist in any given summer, plants living in snowy areas may spend all summer covered with snow. Snow cover, however, is generally protective. Plants beneath it have less exposure to extreme temperatures, less winter desiccation, and less wind and blowing ice. For example, one study showed that a 14 in (35 cm) snow layer in arctic tundra kept the temperature at the top of soil near freezing even though winter air temperature fell as low as $-27°$ F ($-33°$ C). In Siberia, less than 8 in (20 cm) of snow can delay the date soil freezes by two months, even though air temperature may be $-40°$ F ($-40°$ C). The deeper the snow is, the shallower the frozen soil will be. With little or no snow, soil freezes deeply, subjecting plant roots to physical and physiological stress. Some of the coldest air temperatures occur on clear nights just after the passage of a storm. Even a thin layer of snow helps protect plants. During spring snowmelt, however, a snowbank will have a negative effect on soil temperature nearby because heat is used to melt snow rather than warm the soil. Snow accumulations that melt late also shorten the growing season.

### Snow as Insulation

What do a snow cave and a potholder have in common? Why does a potholder help you remove a hot pan from the oven? It is not the thickness of the material, but the air inside. Air is a poor conductor of heat, so the air spaces in the potholder keep the heat in the pan away from your hand. Water is a good conductor of heat, so if the potholder is moist, heat will be transferred in seconds and your hand gets burned.

Fresh snow has up to 10 times more insulating capacity than old, compacted snow because it has a lot of air spaces. With fewer air spaces, old snow conducts more heat from the ground and offers less protection. An ice crust also lowers thermal protection because fewer air spaces exist and heat is more efficiently conducted away from the soil. Because wet snow is a better conductor of heat, it has less insulating capacity. A snow cave will keep you warm by not transmitting your body heat to the outside air.

## Plant Effects on Microclimate

Plants themselves affect and alter their own microclimate, and cold arctic and alpine climates are not always as cold for the plants as official temperature measurements indicate. Plant stature and canopy structure affect air flow and heat exchange, changing the conditions plants experience. Different parts of plants frequently have different temperatures. Leaves at ground level may be 85° F (30° C), while the roots of the same plant are in frozen soil. Short plants closer to the ground usually accumulate more heat than tall plants. Surface temperature near colonies of haircap moss in Antarctica can be 45° F (25° C) higher than that of the surrounding air. Cushion plants, prostrate dwarf shrubs, and herbaceous rosettes (see p. 15 on Growthforms) have leaf temperatures that differ the most from air

Weather data used in climate records are measured at a standardized height of 5 ft (1.5 m) above the ground in a shaded location. A thermometer absorbing solar radiation from direct sunlight would not accurately measure the temperature of the *air*. Ground temperatures are more extreme than air temperatures because the surface absorbs solar radiation during the day and reradiates it at night. The consistency in measurements reduces the effects of the surface and ensures that data from a wide variety of locations worldwide can be compared with no complications.

temperatures. Surface temperatures compared with air temperatures are more pronounced on sunny days and almost negligible under cloudy conditions. Heat accumulates whenever it is sunny. Cushion plants are particularly efficient at trapping that heat. Their smooth, closed canopy inhibits radiation loss. Some plants retain heat by means of hollow stems which may be 36° F (20° C) warmer than the outside air. Fine hairs or fibers hold heat and also shade plants from strong solar radiation. Rosettes and cushions both hold in heat and decrease wind. Warmer conditions in the ground canopy, however, result in a steep temperature gradient between the ground and air, which promotes water loss via more evaporation and transpiration. Root temperatures are also warmer under low-stature vegetation. Deep roots and rhizomes, however, are colder than the shoots.

Tussock grasses, taller shrubs, and krummholz experience temperatures more similar to those measured at standard weather station height and do not undergo extreme diurnal temperature changes. The lower, dead leaves of tussock grasses serve as both insulation and windbreak, keeping the interior growing shoots warm. Measurements in a fescue tussock in East Africa illustrated that temperature of the outer leaves had a large diurnal range, while temperatures at the base varied little. Giant rosettes in tropical alpine environments are warmer than the air because their large leaves absorb heat. They also usually grow in areas with little wind so the heat is not easily dispersed.

Darker leaves, possessing red pigment mixed with green, absorb more solar radiation and have a distinct advantage in absorbing heat under a snow cover. Some plants have thick and waxy leaves (sclerophyllous), enabling them to better withstand wind and blowing ice crystals. Dead leaves and branches often serve as a wind break. Other plants, especially mosses and lichens, are able to absorb water through their leaves, a distinct advantage if soils are dry or soil water is frozen.

## Soils

In general, soils are poorly formed entisols or inceptisols, because most arctic and alpine tundra areas have only recently been free of ice with little time for soil development. The cold climate inhibits chemical actions that release nutrients from parent material such as rock or sediment. Organic matter is limited because of sparse or small vegetation. Soil development is often interrupted by solifluction or other processes related to permafrost. Needle ice is common on moist, fine-textured soils. As surface soil freezes overnight, small ice pedestals lift soil particles and small stones up to 1 in (2.5 cm) above the surface.

However, physical location, parent material, and drainage do cause variation in tundra soils. Rocky arctic areas scraped bare during the Pleistocene remain so, with little or no soil development. Gravelly surfaces are common in the coldest and driest polar deserts. In contrast, sediment overlying permafrost is saturated and boggy. Alpine areas also have extremes, with exposed rocky ridges having little to no soil. Sediment collecting deep between boulders, however, may form a rudimentary soil in small pockets. Former lake beds on valley floors where accumulated sediment supports dense meadows often have initial development of soil horizons. Poorly drained areas form bogs with histosols.

Although not much is known about the nutrient requirements of arctic and alpine plants, nitrogen and phosphorus are the most limiting. Bird and animal waste enriches the soil with nitrogen, stimulating plant growth. Species richness increases around animal dens or burrows and bird nesting grounds. After population explosions of lemmings in the Arctic, surviving plants grow more luxuriantly. Nitrogen enrichment is significant because few nitrogen-fixing plants occur above treeline. Cold temperatures also limit the activity of nitrogen-fixing bacteria in the soil or in root nodules.

Permafrost, a condition in which the subsoil is permanently frozen with only the surface layer thawing in summer, is widespread and may be continuous in the Arctic. Patchy in alpine zones, some areas of permafrost may be relicts of past glaciations. Permafrost may be continuous in mountain areas above the upper level of plant growth. Permafrost not only prevents deep rooting but also locks up nutrients in frozen soil. Melting of the abundant ground ice may disrupt the surface, causing collapse, buckling of vegetation, or soil movement. Destruction of vegetation, road construction, off-road vehicle use, or any activity that alters the heat at the surface can cause the underlying ice to melt. See Chapter 2, Arctic and Antarctic Tundra, for a thorough discussion of permafrost.

## Origins of Arctic and Alpine Flora and Fauna

Although regional variety exists in arctic and alpine floras, so do commonalities. The most important arctic-alpine plant families worldwide are sunflower, bluegrass, mustard, pink, sedge, rose, and buttercup. Also significant and widespread are gentian, carrot, mint, primrose, bellflower, and buckwheat families. Shrub families are primarily heather and sunflower. Representatives of other families are characteristic in parts of the Southern Hemisphere.

Alpine floras are a mixture of plants, including widespread elements, immigrants from adjacent regions, and recent evolution. It is estimated that before the Pleistocene, circumpolar flora included about 1,500 species of vascular plants. Ice sheet advances caused arctic species to migrate south. With glacial retreat, alpine survivors migrated back north to the Arctic. Complex north and south migrations occurred several times, intermixing the arctic-alpine flora. As the climate warmed and cold-adapted species retreated upward in elevation, alpine tundra vegetation became isolated on separate mountains, allowing the evolution of forms distinct to each locale. Glaciation also isolated flora in higher-elevation refugia or nunataks, from which they recolonized when the ice retreated. Low-elevation corridors that retained cool climates for a limited time after glacial episodes also allowed plants to spread. For example, edelweiss, originally native to central Asia, extended its range westward to the Alps. Relicts of Carline thistle remain in the Alps, but the species is now more extensive in Siberia. Because of the east-west orientation of Asian mountains, flora is more fragmented, with more regional variation than in North America. American mountain ranges with a north-south orientation have more similar floras because the continuous mountain chains provided more connected migration routes. Areas that stayed ice-free, such as the north slope of Alaska, central and western Alaska, and the western and central Canadian archipelago in North America, served as refugia. Nonetheless, Pleistocene glaciations and climate changes reduced the circumpolar flora to about 1,000 species.

The ability to cope with environmental changes and extremes demanded evolutionary adaptation. Many arctic-alpine plants evolved from lowland species. The incidence of polyploidy (more than the normal number of chromosomes) increases with latitude. In the low-latitude Arctic, 60% of vascular plants are polyploid, while 70% are so in the high-latitude Arctic. Polyploidy may have provided more genetic variation to enable plants to withstand increasingly severe conditions. Some mountain plants are ecotypes, plants with genetic variations adapted to harsher conditions compared with lowland members of their species. Ecotypes represent an early stage in evolution because they have not yet evolved into different species. Deserts and arctic-alpine plants may have some common ancestors. Both types of plants require physiologies that tolerate dehydration, either because of drought or freezing temperatures. Endemics are plentiful.

........................................................

**Chromosome Numbers**

Cells are normally diploid; they have two sets of chromosomes, one each from the male and female parent. Some organisms, however, are polyploid. They have more (3, 4, 5, or even 6) copies of each chromosome. Triploid means there are three sets, tetraploid means there are four sets, and so on. Polyploidy is common in plants and can occur by natural means or can be induced artificially in a laboratory. Because the plants have more chromosomes, more combinations are possible, and they may have greater adaptive ability.

........................................................

## Plant Adaptations

Arctic and alpine plants grow in stressful environments, with low air and soil temperature, low nutrient availability, extremes of atmospheric and soil moisture, and a short growing season, all of which limit both growth and sexual reproduction. Some plants grow in cold water-saturated soils with little available oxygen, while others grow in habitats that change from wet after snowmelt to dry in mid-summer. Arctic and alpine plants share many adaptations, but some may be more important than others because of variations in local conditions. Arctic soil and air temperatures tend to be lower than those in alpine habitats because of permafrost. Alpine soils are generally better drained because of slopes and less permafrost, so drought stress is more common in alpine environments. However, drought stress is important in polar deserts (extremely cold tundra), where it is also dry. Both arctic and alpine regions can be windy environments, especially on exposed ridges, which usually are dominated by plants of low stature such as cushions and rosettes. Light regimes vary, and plants such as alpine sorrel and spike trisetum may have different ecotypes adapted to continuous summer light in the Arctic or 15–18 hours in mid-latitude alpine situations. Some alpine plants more efficiently use carbon dioxide and have higher rates of photosynthesis because less carbon dioxide is available in the thin air at higher elevations.

**Upper Limit of Plant Growth**

The highest vascular plant in the world, *Saussurea gnaphalodes,* is found on scree at 21,000 ft (6,400 m) on the north side of Mount Everest. Widespread in the Himalayas and other mountains, it is an herbaceous perennial rosette cushion in the sunflower family with thick white hairs. Other plants holding records for life at high elevations are a sandwort (*Arenaria musciformis*) at 20,413 ft (6,222 m) and greater stitchwort (*Stellaria holostea*) at 20,132 ft (6,136 m), both in the Himalayas. Lichens and mosses grow to even higher elevations than vascular plants because they can tolerate complete desiccation. They are found at 19,350 ft (5,900 m) on Mount Kilimanjaro, 22,000 ft (6,700 m) in the Andes, and 24,250 ft (7,400 m) in the Himalayas. More than 50 species of lichens grow above 13,100 ft (4,000 m) in the European Alps.

## Growthforms

Chamaephytes, plants that hold their regenerating buds just above soil level, are the most common growthform in arctic and alpine biomes. Hemicryptophytes, which hold buds at the surface, are also common, but they are less important in alpine tundra than in arctic environments. The two most common growthforms in alpine tundra are short or prostrate woody shrubs and herbaceous perennials of several types. Less common or regional growthforms include giant rosettes on tropical mountains, geophytes on mid-latitude mountains with pronounced seasonality, stem and leaf succulents, and annuals or biannuals, which become rarer at higher elevations. Although most arctic plants use the $C_3$ pathway in photosynthesis, they can usually photosynthesize at lower temperatures than their temperate counterparts. Succulents, which are capable of the Crassulacean Acid Metabolism (CAM) method of photosynthesis, live in dry niches at all latitudes, including subarctic. Bryophytes and lichens, which are nonflowering and often desiccation tolerant, are common and often extend to higher elevations. Because alpine tundra is defined as above treeline, few or no phanerophytes (plants such as trees that hold buds high above the ground) occur. Some growthform characteristics may be due to poor nutrition as well as to low temperatures.

Prostrate shrubs are well adapted to arctic-alpine conditions. The fact that arctic willow is found north of 83° N, almost to the limit of plant growth in the Arctic, indicates its hardiness and resistance to cold conditions. A major advantage of the small shrub growthform is its permanence because new tissue is not needed every

**Cryptogams and Phanerogams**

The term cryptogam, literally meaning hidden seed, refers to various algae, fungi, mosses, lichens, ferns, and other lower plant or plant-like organisms that lack flowers and reproduce by means of spores too small to be seen. Blue-green bacteria, green algae, slime molds, fungi, lichens, molds, and yeasts are in a category called thallophytes. Liverworts and mosses are bryophytes, structurally simple green, seedless plants. Many cryptogams, especially algae, are tiny and must be observed with a hand lens or microscope. Most others are small, 0.8–2 in (2–5 cm) tall or less than 4 in (10 cm) long. They form flattened mats, spongy carpets, tufts, or turfs according to humidity and sunlight. Most cryptogams are nonvascular plants. Although categorized as cryptogams because they do not produce seeds, club mosses, horse-tails, and ferns are vascular plants called pteridophytes. Cryptogams form a significant part of arctic-alpine floras.

In contrast, the term phanerogam refers to plants with visible seeds, like flowering plants such as angiosperms and gymnosperms. Vascular plants such as phanerogams have differentiated tissue designed to transport water and nutrients from roots to leaves.

**Lichens**

Lichens are organisms composed of a symbiotic association between fungi and algae that takes several forms. Crustose lichens are crust-like, growing flat against the rock or soil. Foliose lichens are leaflike, and fruticose lichens are shrubby or hair-like. In addition to green, they can be yellow, orange, or brown, with many color variations.

year. The plant can wait out bad years before resuming growth and reproduction. This advantage is even more significant if the shrub is evergreen. Ground-hugging or prostrate plants are covered and protected by winter snow. They are also able to absorb more heat from the ground in summer.

Plants may be evergreen or deciduous. Some, like arctic bell heather or moss plant, withdraw chlorophyll from their leaves at the end of the growing season, leaving winter foliage red. Others remain green. Evergreen plants do not replace their leaves every year and need fewer nutrients; they grow on acidic, low-nutrient soils. Deciduous willows and birches grow more quickly but also need better soil. Several shrubs, especially in New Zealand, transfer carbohydrates and nutrients from leaves to the woody parts of the plants for winter, returning them to the growing buds in spring. Plants that maintain a few leaves above ground begin photosynthesis before snow fully melts because radiation can penetrate snow up to 20 in (50 cm).

The most typical tundra growthform, however, is the herbaceous perennial, which depends on long roots and rhizomes. Four major types—graminoid, cushion forb, mat forb, and leafy forb—dominate along with a few small ferns and bulbs (see Figures 1.2 and 1.3). Most have a deep root system that serves both to store carbohydrates and extract deeper soil moisture. Surface roots absorb heat from the soil layer that warms quickest, while large tap roots store energy and nutrients. Thick taproots may stabilize plants under frost-heaving or needle-ice conditions, but, thick roots may break when frozen.

Graminoid plants include both grasses and sedges, usually growing in clumps or tussocks. Grasses have hollow round stems, while sedges are usually triangular in shape and have a noticeable difference to the touch.

Low-growing with a streamlined shape, cushion plants such as the circumpolar moss campion,

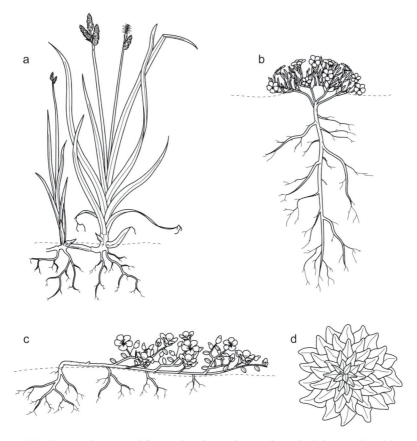

**Figure 1.2** Four major growthforms dominate the arctic and alpine tundra: (a) graminoid, (b) cushion, (c) mat, and (d) leafy forb. *(Illustration by Jeff Dixon. Adapted from Zwinger and Willard 1972.)*

are common pioneering plants in exposed, windy areas. Formed of many short branches that fit tightly together, they are not soft like the name cushion implies but present a firm rounded surface. The plant may be sculpted by wind and ice because any twig projecting above the general level is quickly cut back. The tightly packed branches expose a lot of leaf area at the cushion top, but beneath the canopy, it is warm and protected from the wind. Cushion plants improve their microhabitat by accumulating plant debris and nutrients beneath their canopy. Moss campion grows in size by adding more branches, but growth rates are slow. Only 0.5 in (1.3 cm) in diameter after the first five years, the plant may need 25 years to reach 7 in (17.8 cm). Initial growth is slow because energy is directed toward development of a taproot system. A 10 in (25.4 cm) diameter plant can have a 4–5 ft (1.2–1.5 m) taproot, good for both anchoring the plant and obtaining subsurface water. Cushion plants in South America and New Zealand, however, grow much larger.

**Figure 1.3** Four major growthforms of arctic and alpine tundra include (clockwise from top left) sedge, moss campion, alpine sandwort, and big-rooted spring beauty. *(Photos by author.)*

Mat plants, such as dwarf clover and alpine sandwort, are a looser tangle of prostrate branches that root over a large area. Ground-level branches and short upright stems are not so tightly packed as in cushions and are not dependent on one long taproot. The plant grows larger as prostrate branches root where they touch the ground. Because they have a short structure of woody twigs, cushions and mats are often called tiny shrubs.

Rosettes, such as tundra dandelion, are a major type of leafy forb. They grow flat to the ground, sometimes closer to the ground than cushions or mats, in the warmest microenvironments. Overlapping leaves, which form a circular pattern on a very short stem, present maximum exposure of leaves to light. Smaller rosettes are offset from the mother plant. Some plants, especially rosettes such as snowball saxifrage, big-rooted springbeauty, and yellow stonecrop, have succulent leaves or a thick taproot. The taproot of big-rooted springbeauty can be 6 ft (1.8 m) long or more, while the succulent leaves of stonecrop form marble-size balls. Waxy leaf surfaces contribute to moisture retention.

Plant hairs both offer protection from intense radiation and drying wind and also help to absorb or conserve heat (see Figure 1.4). Dark hairs such as found on draba, fleabane, and arctic poppies are less reflective and absorb more solar radiation. In contrast, light-colored, silky, wooly, or transparent hairs such as on pasqueflower, some lousewort, cottongrass, and some willow catkins allow solar radiation to pass through where it is then absorbed by the darker surface of the plant. Longwave energy reradiated from the plant cannot escape and the energy is trapped in a mini-greenhouse of hairs. The fluffy tops of arctic cottongrass tussocks retain heat that delays soil freezing. Freezing of the surrounding soil pushes the mound of cottongrass upward where it occupies a position to get more of the sun's rays in spring. It emerges from snow before adjacent, slightly lower-lying plants do, increasing its growing season by 4–10 days. Some of the hairiest plants

**Figure 1.4** Like pasqueflower, many arctic and alpine plants have hairs to both protect from excessive sun and retain heat. *(Photo by author.)*

are the earliest to flower because they trap more energy or heat. Hairiness takes many different forms.

## Coping with Climatic Stress

Both arctic and alpine plants are subjected to extremes of low temperature. Plants are more susceptible when they are not frost hardened. During winter, the plants' physiology alters, usually through dormancy, so cold can be endured. A decrease in the photoperiod (hours of daylight) and lower temperatures at the end of summer also harden plants. Soil drought may be another factor in winterhardening for alpine plants. From spring to fall, and every night in the Tropics, growing tissue is more susceptible to freezing. Aboveground tissue, especially new growth, may die, but the plant is usually not killed. If new growth has not yet begun, or if the plant is still protected under snow, it will not be affected.

Tolerance limits of alpine plants vary with geographic location and plant parts. The stem and root are usually more tolerant of low temperatures. Low-stature plants (cushions, dwarf shrubs, rosettes, and sedges) have similar tolerance limits regardless of family or genera relationships. Mid-latitude alpine plants in the temperate zone such as the Alps can tolerate 14° to 25° F (−10° to −4° C) without damage. Earlier flowering plants, which are more likely to be exposed to colder early summer temperatures, are most cold hardy. Tropical alpine plants in the Andes, Africa, and Hawaii can tolerate temperatures even lower, −2° to 16° F (−9° to −19° C). A more stable temperature during the growing season is responsible for the higher tolerance limits in the mid-latitudes. Low temperatures in the Tropics are less predictable, which explains the lower tolerance limits.

Plants have several ways of coping with low temperatures. The first is to avoid exposure. In nontropical mountains with seasons, phenology (the date of growth activity such as flowering and setting seed) is important. Plants are dormant in winter in spite of periodic warm spells and potentially high radiation and cannot be damaged by extreme cold. With warming temperatures and longer daylight hours in spring, plants become active and are most sensitive to cold in summer. A decreasing photoperiod in late summer signals the plant to finish its growth cycle before fall. As nutrients in the leaves are reabsorbed by the roots or stems when the leafy parts die back, the plant's hardiness to cold is increased. Plants can avoid exposure to cold by their morphology, particularly plant size and position of regenerating buds. Tall plants are above the snow cover and exposed, while short ones are protected from air temperatures. Most tundra plants carry their regenerating buds at or close to ground level. Graminoids and perennial forbs, both of which are abundant in arctic and alpine environments, carry regenerating buds below ground level, pulled back into the soil as roots contract. Cushion plants store heat under their low canopy. Tropical rosettes curl their leaves over the center growing point, and dead leaves insulate the stalk. Liquid inside the rosettes keeps the plants warm. Microhabitat preference (open versus sheltered, snowbeds versus bare) is a simple way to avoid exposure.

Many plants have the ability to avoid freezing even when temperatures are low. By accumulating sugars in tissues, the freezing point is lowered, not by much, however, and this adjustment is least effective. A more efficient method is supercooling. In a physiological process, the leaves and stem cool to below-freezing temperatures, but tissues do not freeze and suffer damage. Common in areas where low temperatures regularly drop to $10.5°$ F $(-12°$ C), such as encountered in the South American páramo, this method is used by the giant rosette *Espeletia* and some shrubs. If temperatures drop below the limit of supercooling, however, tissue freezes quickly and damage occurs. Giant rosettes in tropical alpine areas in Africa experience even lower temperatures but do not supercool, nor do higher-elevation Andean plants.

Some plants are able to tolerate freezing. In all plants, 25–30% of the plant is space between cells. Water in intercellular spaces (that is, not inside the cells) freezes, and in the process, releases heat to the cells and takes water from the cells. Water or ice between the cells makes the leaf appear darker in color. When thawed, the cells reabsorb the water within one to four hours and the normal green color returns. Some plants can begin growth in spring, then cease growth and wait out temperatures below freezing at no harm to the plant until conditions improve.

Plants adjust to the cool temperatures of the short growing season by preforming flower buds at the end of the summer. Growth of buds and flowers usually takes more than one season. Catkins of some prostrate willow species develop over four growing seasons compared with two seasons for shrubby species in better environments and one year for lowland species. Protected by the plant and snow cover over the winter, buds quickly elongate and flower early the following season. Some plants bloom just two weeks after beginning to grow in spring. Solar radiation can penetrate up to 2.0 in (50 cm) of snow, an important factor for buttercup, which begins growth before the snow melts. However, the upper layer of soil must be thawed for growth to take place. Preformed flower buds are especially common in New Zealand alpine tundra. Spring growth, beginning with elongation of preformed leaves and flowers, depends on the supply of nutrients from storage organs as well as increasing receipt of solar energy. Roots, rhizomes, corms, and bulbs store energy and nutrients. The root-to-shoot ratio increases with both latitude and elevation. Root organs become larger as the aboveground part of the plant becomes smaller, resulting in more stored energy. Old leaves serve a similar purpose in small shrubs, recycling the nutrients, but even prostrate shrubs are absent from the harshest climates.

A rare adaptation in arctic-alpine plants is to repair or replace tissue damaged or killed by freezing. Preformed leaf buds and flower buds lost

**Heat Resistance**

Because surface temperatures can be extreme, tolerance of some alpine plants to high temperatures is similar to plants found in tropical deserts. Lethal temperatures of leaf surfaces (that is, temperatures at which plants die) in the Central Alps is $112°$ F $(44°$ C), the same as the average in Mauretania in the Sahara. In the Hindu Kush mountains, the lethal limit is $102°$ F $(39°$ C), and in central Norway, $106°$ F $(41°$ C).

to frost will not be replaced because the growing season is too short. For the same reason, flowers or fruit frozen later in the summer cannot be replaced. Fat roots, however, do store nutrients and to some extent can replace frozen or eaten leaves. Graminoids and rosettes can activate belowground growing points if damage is not extensive.

Alpine plants are also subject to heat stress. Heat-trapping structures can be detrimental during peak solar radiation, causing ground and plant temperatures to soar. Establishment of seedlings where high surface temperatures prevail is difficult at best and is another indication of why vegetative propagation is more common. Patches of bare soil are frequent in alpine regions, and seedlings are commonly found in disturbed sites, such as animal burrow entrances and tire tracks. Any activity that destroys vegetation opens the ground to accumulation of surface heat. However, such bare spots can be permanently damaged by erosion before they can be reclaimed by vegetation.

Surface soils, particularly in alpine environments, can become quite dry, but because deeper moisture is often available, plants have few adaptations related to tissue desiccation. Plants showing drought stress usually grow on exposed sites or in shallow soil. Drought stress in the high-latitude Arctic is common in some grasses, rosette species, and cushion plants in bare, well-drained soils.

To gain the most energy from solar radiation, some plants orient their leaves vertically to catch the low sun at a more efficient angle. This vertical orientation also allows light to be reflected to and absorbed by other leaves instead of back to space. The parabolic form of several flowers is beneficial for both pollination and development of seeds. The white and yellow flowers of avens, poppy, and buttercup have good reflectivity and concentrate energy toward the flower's center. Some flower heads maintain an orientation toward the sun like a lowland sunflower. Many alpine plants not considered evergreen retain some green color in winter. Examples include some leaves of cinquefoil and matgrass. The bark of young shoots of dwarf shrubs, especially the green stems of blueberry, can conduct photosynthesis. Mosses and lichens undertake photosynthesis in winter, and all alpine plants are capable of photosynthesis processes at temperatures near freezing. While much solar radiation is reflected from snow, some wavelengths penetrate, depending on snow's thickness. If snow is shallow, some, but not all, alpine plants can carry on photosynthesis.

Lichens and some mosses can withstand colder or shorter growing seasons, but they need water. Lichens have advantages in the low stature of their lifeform, close to warm daytime surface or rock temperatures, and those warmer temperatures may melt snow needed for water. Their optimum photosynthesis temperature is about 41° F (5° C), a temperature easily reached at the surface. Low night or winter temperatures do not harm lichens, and they can remain dormant while frozen. They are also long lived and can survive unfavorable conditions in a dormant state. In the Alps, map lichen lives up to 1,300 years, and in western Greenland, up to 4,500 years. Lichens grow beyond the limits of vascular plants but eventually are limited by the lack of water from snowmelt due to low temperatures and sublimation.

## Reproduction

About 90% of the vegetation is perennial, including grasses, sedges, flowering plants, mosses, lichens, and prostrate shrubs. Age estimates based on growth rate of leaves and rhizomes suggest that many arctic species are long lived, delaying the need to reproduce either sexually or vegetatively. Stems of birch and willow may be 200–400 years old, and dwarf birches rising only a foot high may be hundreds of years old. Other plants may have equally long life spans: northern woodrush, 90–130 years; avens, 80–120 years; and cottongrass, 120–190 years.

While the most common method of reproduction is vegetative rather than by seed, most arctic-alpine plants reproduce both ways. Flowering and seed production, however, may not be successful every year. Most plants are insect pollinated, but wind and birds are also factors. Long-tubed flowers are usually pollinated by bumblebees, which are attracted to colorful flowers. No long-tongued bees such as bumblebees are native to New Zealand alpine tundra, and unlike the mid-latitudes and the Arctic, which both have colorful flower displays, the mostly small, flat, white or yellow flowers in New Zealand are pollinated by small insects, flies, and short-tongued bees.

Some plants are biennial, taking two years to flower, but few are annuals because of the short growing season. In Peary Land in Greenland, only 1 of 96 vascular plants, the Iceland purslane, is annual. In the alpine environment of the Rocky Mountains in Wyoming, only 3 of 121 plants are annual. In general, annuals comprise only 1–2% of the flora in alpine or arctic areas and plants are typically small. Drier alpine regions have slightly more annuals, for example, 6 of 108 species in the Sierra Nevada in California. Some species, such as rock primrose, that are normally annuals have perennial ecotypes in arctic or alpine tundra.

Arctic annuals have a low root-to-shoot ratio with no need to store energy because the plant only lives one season. Their growth is indeterminate, meaning that they will continue to grow flowers and produce seed until killed by either cold or drought. They use water sparingly, regulating their stomata as the soil dries, and have a high rate of photosynthesis that allows them to complete seed ripening. No annuals are endemic to the Arctic.

Germination of most plants takes place soon after snowmelt when soils have warmed and water is sufficient. Seed dormancy requirements for chilling, light, or seed coat scarification are present in only 20–40% of arctic-alpine species.

An additional safeguard for most plants is that not all seeds germinate every year, a survival strategy that ensures that a disastrous summer season will not destroy the species. Seeds may remain viable for many years. Arctic lupine seeds that were found frozen and preserved for 10,000 years germinated within 48 hours after being thawed and moistened. Seedlings are rare in the Arctic, but Bigelow's sedge and cottongrass are an exception and can be seen in abundance on bare soils in northern Alaska. Seeds, distributed by wind or animals, are usually small with no burrs or fleshy fruits.

Clonal growth is supplemental to sexual reproduction, and few plants rely solely on vegetative means. Most alpine plants in western North America reproduce by seed, but vegetative reproduction is more common in the Arctic because of the difficulty of setting seed there. Alpine sorrel has two ecotypes, one reproducing by seed in alpine tundra, the other by rhizomes in the Arctic. Plants which reproduce sexually at lower elevations may rely on vegetative means in alpine tundra. Vegetative reproduction is also common where wet areas curtail the growing season. Percentage of species with some form of cloning varies from 50–80% of total flora in both arctic and alpine environments, but it may be less important in alpine tundra.

Cloning takes several forms. Many tussock-forming grasses, sedges, and rushes form dense clusters. As new shoots grow at the edges, old ones in the center die. Over time, the tussock becomes a ring of living plants surrounding dead centers or a sinuous pattern of lines (see Figure 1.5). As the most important clonal strategy in the world, it is used by carex and kobresia sedges, rushes, hairgrass, fescue, matgrass, bluegrass, and needlegrass. Some graminoids and forbs grow new plants from stolons or rhizomes that extend either below or above ground away from the mother plant. The invasive nature of mint plants and Bermuda grass in a home garden illustrates the efficiency of reproduction by rhizomes.

Young plants can become completely separated from the parent by rock or soil movement. Graminoid examples include carex sedges, woodrush, bentgrass, and fescue. Examples of forbs that clone this way include primroses, avens, snow lotus, ragworts, dock, bellflower, and candytuft. Some mat-forming forbs such as pussytoes, *Celmisia,* and strawflower form herbaceous clusters of rosettes. As plant

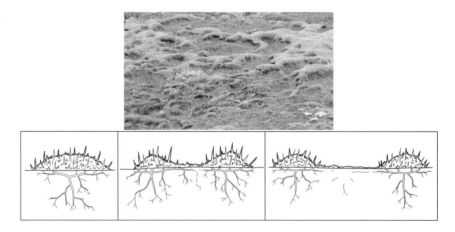

**Figure 1.5** Because the center dies as the edges expand with growth, vegetative reproduction of many types of grass tussocks produces rings or a sinuous pattern of living plants. *(Illustration by Jeff Dixon. Photo courtesy of Susan L. Woodward.)*

clusters grow larger, they disintegrate into separate plants. Other examples are gentians, plantains, and speedwell.

Dwarf shrubs, which are small, close to the ground, and buried in litter, may grow adventitious roots from the woody stems, providing the soil is moist. Typical plants include *Coprosma,* avens, azalea, and willows. In a similar manner, woody stems of taller prostrate shrubs (2–20 in, 5–50 cm, high) may also become buried in litter or soil. They first grow shoots from buried buds, with adventitious roots developing much later. Common plants are crowberry, rhododendron, blueberry, *Hebe,* St. John's wort, willow, and *Styphelia.*

Viviparous plants form vegetative propagules on inflorescences. A subspecies of alpine bluegrass produces plantlets, and viviparous bistort produces bulbils (seed-like shoots) instead of flowers. Some plants that do not ordinarily reproduce vegetatively, such as glacier buttercup and alpine sorrel, sometimes grow roots from flower stalks that become accidentally buried.

Many lichens reproduce by means of fragmentation. Broken pieces subsequently root to create clones of the parent. As it dries, staghorn lichen contracts into a ball form that is blown around the tundra by the wind, breaking off pieces as it rolls. With moisture, it unfurls and continues growth.

## Animal Life

Arctic and alpine environments have distinct assemblages of animals. With the exception of high-latitude mountains that merge with arctic tundra in northern Alaska, northern Urals, and northeastern Siberia, no mammals and few birds are common to both regions. Many animals in the Arctic have a circumpolar distribution, meaning they are found in all or most regions surrounding the North Pole. Alpine animals, however, which derive from adjacent lowlands rather than from the Arctic, have patchy distributions because the alpine areas of high mountains are like islands in a sea of unsuitable habitat. Isolated areas of alpine environment are more difficult for animals to colonize than the continuous expanse of the arctic tundra.

Animal life of arctic and alpine regions can be categorized as permanent residents, seasonal breeding visitors, or seasonal but nonbreeding visitors. Patterns of distribution can be continuous, disjunct, or disperse. Distribution areas of terrestrial arctic birds and mammals are usually continuous, especially on a given continent, although each species will occupy only a particular habitat within that large region. Lapland Long-spur, for example, is found in both North America and Eurasia but not over the oceans. In contrast, marine birds and mammals, such as Arctic Skua, Black Guillemot, and ringed seal, are found on both the oceans and continents, with seals being limited to coastal areas. A continuum exists between the terms disjunct and disperse, with disjunct referring to broad distribution areas that are widely separated, such as caribou in Canada and Alaska and reindeer in

Fennoscandia. Disperse distributions refer to widely scattered areas, such as marmots or pikas limited to mountain tops in mid-latitudes.

Alpine regions of the high plateaus of Central Asia have the largest variety of animals. Because this is the richest tundra, it is believed to be the origin of animal life now found in the Arctic. During the Pleistocene, while most of North America and Europe was covered in ice, northeastern Siberia was not because the climate was too dry to accumulate enough snow. Animals thrived in this cold, but ice-free zone. As the Bering Strait land bridge emerged due to dropping sea level during the Pleistocene, several cold-adapted animals moved across to Alaska, where they could then spread throughout North America. Along with extinct animals, such as wooly mammoth and sabre-tooth cats, North America gained several modern species, including moose, lemming, hare, muskox, fox, and caribou from Eurasia.

## Adaptations to Cold

***Morphological or physiological adaptations.*** Tundra animals must be adapted to a short season of plant growth and to a cold winter too long for hibernation. The most obvious adaptation in these cold environments is the need to regulate body temperature by maintaining a balance between heat loss and heat production. Most animals have a small surface-to-volume ratio, which minimizes heat loss, as do short appendages such as ears and limbs. Arctic fox, for example, can tolerate temperatures of $-58°$ F ($-50°$ C). Many animals, such as muskox and brown bear, have a thick layer of fat beneath the skin that serves as insulation and stored energy. Mammals often have a double coat of fur, one of which is a dense and fine undercoat for insulation. The amount of heat lost depends on the steepness of the temperature gradient between the animal's warm body and the cold air. Warm skin will radiate and conduct heat quickly, but an insulating layer of fur or feathers will lower the temperature gradient, allowing the animal to retain more heat. Fur and feathers along with air trapped in the pelt are poor conductors of heat, so with no wind to disturb the insulating layer, the animal stays warm. Because an efficient winter coat would trap too much heat in summer, many birds and mammals undergo a seasonal change. Mammals like muskox have longer or thicker fur in winter. Bird feathers do not change in terms of thickness of plumage, but the structure of the feathers traps more heat in winter. Muscular contractions that fluff up feathers increase the thickness of insulation, but thick fur on mammals cannot be so easily fluffed.

...............................................

**Ptarmigan Camouflage**

Seasonal color changes help camouflage both birds and animals. Along with the weasel (called ermine in winter), arctic fox, and arctic hare, the ptarmigam changes to white in winter (see Plate III). The three species of ptarmigan have three seasonal plumages, changing not only color but also thermal properties from winter, to spring, to autumn. The loss of pigment in the white feathers leaves air spaces that trap heat. The birds' legs are feathered for warmth, and feathers around their feet in winter act as snowshoes. Because the female sits on the nest first, she is also the first to change to speckled camouflage plumage in spring.

...............................................

Animals may control the amount of metabolic heat their bodies produce by increased activity such as running, digging, or shivering. Ptarmigan feathers, however, provide such good insulation that the birds have no need to shiver. Comparing the same thickness, feathers are better than fur for insulation, so mammals must be larger to produce enough metabolic activity to keep warm. Arctic fox and arctic hare are the smallest animals to survive with just their basal metabolism, but they are 10 times the weight of ptarmigan. Some birds go into torpor by reducing their body temperature at night, which in turn decreases their need for energy by 25–50% of normal.

Many body extremities have either no insulation or are covered with thin hair. Examples include nostrils, toe pads of arctic foxes, soles of brown bear and polar bear feet, hooves of caribou and bighorn sheep, and feet and legs of most birds. To avoid losing heat through these areas, a physiological adaptation controls the volume and temperature of blood flow. Constriction of blood vessels limits flow and allows these extremities to become cooler (see Figure 1.6). An exchange of heat also takes place from arterial blood to venous blood. Blood in arteries moving toward extremities transfers heat to veins transporting blood to the body core. Therefore, heat is retained in the body and cooled blood is sent to the extremities. Gulls' feet, for example, can be close to freezing with no harm, thereby lowering the temperature gradient between the animal and the icy surface and limiting heat loss.

During the warmer summer or periods of increased metabolic activity such as running, animals may need to expel extra heat. Blood vessels then dilate and the internal heat exchange process is bypassed so excess heat can be taken to and lost through the extremities. Another major means is by panting, used by canines and even by ptarmigan. Evaporative cooling through sweat or saliva on the fur or feathers is not used because the moisture would interfere with the insulating qualities of the pelt.

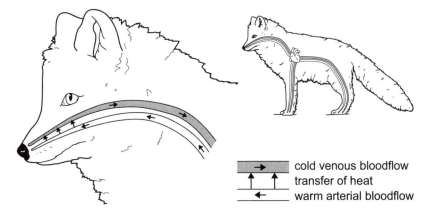

cold venous bloodflow
transfer of heat
warm arterial bloodflow

**Figure 1.6** A system of heat exchange between arteries and veins helps animals in arctic or alpine environments maintain body temperatures. *(Illustration by Jeff Dixon.)*

*Behavioral adaptations.* Some animals depend on behavior to keep warm. Small animals have neither a thick fur coat nor a high enough metabolism and must avoid exposure to cold conditions. In true hibernation, body temperature decreases until it approximates that of the shelter, and with little to no temperature gradient, the animal loses little to no heat or energy. Several marmots and ground squirrels in alpine tundra hibernate, but only one arctic species does, the arctic ground squirrel in eastern Siberia and western North America. Arctic environments, especially with permafrost, offer few places to find a shelter that will not freeze, and the arctic ground squirrel is restricted to well-drained sandy soils well covered by snow. The burrow entrance is below the level of the chamber, so heat is trapped as it rises from sleeping bodies. Because the soil is well drained, burrows and sleeping chambers are dry with no snow or ice to steal heat for either melting or evaporation. As a result, temperatures do not drop below $10°$ F $(-12°$ C) regardless of outside conditions. Although not true hibernators, bears reduce their metabolism as they den up for winter. Some mammals, such as chipmunks and hamsters, are torpid but wake up periodically to eat. Others, such as deer mice and pikas, are active all winter and cannot become torpid.

Many rodents, such as lemmings, avoid winter cold by maintaining active life in tunnels beneath the snow where they continue to feed on roots, stems, and buds. Animals that do not hibernate must have food to carry them through the winter. Pikas, deer mice, wood rats, Old World hamsters, and narrow-skulled voles store food, not fat, for the winter. They live in insulated nests on or above the ground or sometimes in shallow burrows. Piles of grass left to dry on rocks by pikas before storage are common sights in North American mountains. Lemmings do not store food but continue normal foraging behavior beneath an insulating layer of snow where temperatures are not much lower than freezing and nests are lined with grass and old fur.

Arctic and alpine animals have similar adaptations to cold although they may not be taxonomically related. Cold steppe and desert animals in the mid-latitudes have many of the same adaptations to cold winters, and it would have been relatively easy for them to adapt to alpine conditions.

## Role of Snow Cover

Snow cover can be either a benefit or detriment depending on the depth and the animal. Many small animals need an insulating snow cover in winter, with each species (such as voles, lemmings, deer mice, and pikas) finding it in a preferred habitat (such as depressions, krummholz, or patterned ground). Snow-free areas are uninhabited. Distributions and behavior of large animals are also related to snow cover. Snow in the Arctic is reworked by wind into mobile drifts or a crusty cover. While drifting snow may uncover vegetation, herbivores most often must dig through snow to forage. If snow is too crusty, caribou migrate to softer snow at the edge of the forest, but in areas of soft snow, animals do not need to migrate unless it is too deep for easy movement. Others herbivores such as mountain goats in mid-latitude alpine environments browse on subalpine trees and krummholz branches that project above the snow cover. Mountain sheep either paw through the snow or graze on ridges that are blown free of snow. Ptarmigan, one of the few winter bird residents, can burrow into soft snow or migrate to taller willows for food or shelter. Most predatory birds migrate, but Snowy Owls hunt where snow is thin or at rodent air holes through the snow. Mammal predators such as wolves migrate with the prey, but arctic fox and least weasel remain on the tundra to hunt the occasional rodent.

## Reproduction
Because most mammals, unlike birds, in both the arctic and alpine environments are nonmigratory, or only travel a short distance like caribou to the forest edges, the short growing season plays only an indirect role in reproduction. Reproduction is more closely controlled by quantity and quality of food resources. Large animals with long gestation periods breed only once a year, and time of birth is correlated with the season with maximum forage. Small mammals such as rodents with short gestation periods have more than one litter during the short summer breeding season. Both the number and size of litters are related to quality of diet, which is also correlated with length of growing season. Some small animals, such as brown lemmings in arctic tundra, which breed under snow all winter, are a major exception. Such breeding does not take place every year but the reason why is as yet unknown. The montane vole in alpine areas breeds beneath snow in the early spring as well as in summer, but probably does not do so all winter.

## Birds
Because of the vast expanse of permafrost and boggy ground in the Arctic, water birds and shore birds (ducks, geese, swans, plovers, and sandpipers) are numerous. No such habitat exists in alpine environments, either mid-latitude or tropical.

In both arctic and alpine environments, many migrant breeding birds make the most of the short summer season by arriving on the arctic tundra or alpine habitat already paired. In other species, such as the Pectoral Sandpiper in the Arctic, courtship time is shortened. In the alpine zone, Rosy Finches and White-tailed Ptarmigan pair after arrival, but Water Pipits may already have mates. Unlike birds in climates with a longer summer season, pairs attempt to nest only once, but clutches are usually larger than birds of more temperate climates. For most passerines in both arctic and alpine environments, the female incubates the eggs but both male and female feed the young. For other birds, males, females, or both may share duties of caring for eggs and chicks. Most birds undergo molting after breeding season is over, avoiding a double expenditure of energy.

## Invertebrates
Adaptations of insects to alpine environments have parallels in the Arctic and Antarctic. Both morphological and behavioral adaptations enable them, like plants, to

survive cold winters but grow and reproduce in short, cool summers. Insects have a variety of ways of surviving long cold winters that last eight to nine months in the Arctic and northern mountains. Some live in or adjacent to unfrozen lakes and rivers where the temperature is around freezing and snow cover protects them from cold. Many rely on supercooling, meaning that the temperature at which they freeze is decreased. An insect's freezing point of 17.5° F (−8° C) in the summer may adjust as temperatures drop in the fall, reaching −30° F (−35° C) in mid-winter. The process reverses as temperatures warm in the spring. Other insects are tolerant of freezing, can survive frozen ice in their tissues, and are able to survive temperatures 23° to −30° F (−5° to −35° C), depending on the species, in a frozen state. When the upper soil layer freezes to an ice cover in winter, insects enveloped in the ice become oxygen deprived but can survive several weeks in an anaerobic state. This mechanism is used by insects living in ponds or lakes that freeze solid.

Morphological adaptations include small size, making it easier for those insects to find enough food and shelter. The *Colias* species of butterflies are largest in the subtropics and smallest in the Rocky Mountains, Himalayas, and Arctic. Pierine butterflies in the Andes are also small with smaller wings than lowland species. They hide in vegetation on the Altiplano to escape low temperatures and windy conditions. The percentage of species having small wings, or none, increases with elevation. About 60% of insects above 13,200 ft (4,000 m) in the Himalayas have small wings, and several are unable to fly at all, compared with lowland counterparts with normal wings. Mount Kilimanjaro in tropical Africa has a similar insect component. Small wings reduce the risk of being carried off by wind. More dark colored insects, which absorb more heat energy, occur at higher elevations. Red-brown is better than black in raising temperature because it more efficiently absorbs infrared radiation, while also protecting from excessive ultraviolet radiation. Alpine butterflies typically have darker bodies and wings than lowland relatives, because they need to absorb more heat energy in order to fly. Similar patterns of temperature regulation are found in alpine blowflies in the Eastern Pamir Mountains and grasshoppers in the North American Rockies.

Predatory insects such as carabid beetles and some spiders that are insensitive to cold are nocturnal, preying at night on insects immobilized by low temperatures. In dry environments, insects avoid desiccation by sheltering in crevices and vegetation. Some grasshoppers in the Andes and Rockies, however, are equivalent to desert insects in resistance to aridity and desiccation. Similar to how plants may need several seasons to develop flowers and buds, some insects take two to four years to complete their life cycles because summer is short. Overwintering can take place at different stages in development. Conversely, some insects, perhaps because of more productive habitat, complete their life cycle in just one year. Several types overwinter as adults, ensuring survival of the species.

Small animals like spiders can be found above the limit of plant growth, living on wind-blown debris in the aeolian zone. Bacteria can also survive if wind-blown organic matter is available.

Amphibians and reptiles are rare in both arctic and alpine environments. In North America, the wood frog is the only amphibian even on the fringes of the Arctic, while only a few toads and the sagebrush lizard can be found in mid-latitude mountains. The Old World has few to none of these species in the Arctic, but more occur in Palearctic alpine environments.

## Human Impact

Whether arctic or alpine, tundra vegetation is easily damaged, and because of its very slow growth, it is also slow to recover. Arctic areas are threatened by oil and mineral exploration. Resulting development disrupts animal migration routes, and oil and chemical spills can be devastating. Permafrost presents problems for construction. Eurasian tundra is further degraded by overgrazing by domestic reindeer. Much of Antarctica's ice-free land is used for research stations with buildings, airfields, garbage, trampling, and noise. In some countries, alpine regions are subjected to heavy grazing, and woody shrubs at the edges of the alpine zone are often burned in an attempt to create more pasture. In some areas, it is hard to determine what the natural vegetation is. The rising popularity of ecotourism may be beneficial in preservation of both arctic and alpine ecosystems, but it requires careful management unless increased numbers of visitors disturb the natural landscape the travelers have come to see.

## Further Readings

Benders-Hyde, Elizabeth M. n.d. http://www.blueplanetbiomes.org.
Encyclopedia of Earth. n.d. http://www.eoearth.org.
Palomar College. n.d. Wayne's Word. http://waynesword.palomar.edu/index.htm.
Weather Base. n.d. http://www.weatherbase.com.
World Wildlife Fund. n.d. http://www.worldwildlife.org/wildworld/profiles/terrestrial_na .html.

# 2

## Arctic and Antarctic Tundra

Tundra refers to several types of vegetation of different low-stature growthforms, including tall shrubs (6–16 ft, 2–5 m), dwarf-shrub heath (2–8 in, 5–20 cm), and graminoids and cryptogams (both 4–20 in, 10–50 cm). Plants, including vascular plants, lichens, and mosses, may cover 80–100% of the ground. The term polar desert (or semidesert) is used to describe extremely barren tundra. Vegetation includes dwarf shrubs, sedges, perennial forbs, mosses, and lichens. Tundra environments are located where climate conditions are too severe for growth of trees.

### Polar Comparisons

Although one might expect both northern and southern polar regions to be similar in terms of climate and environmental conditions, several major differences exist. The Arctic is an ocean almost surrounded by large landmasses, while the Antarctic is a high, ice-covered continent isolated from other continents by cold water which is covered by ice in winter. Sea ice in the Arctic is reduced by one-half in the summer, but by only one-seventh in the Antarctic. Northern Hemisphere tundra lies generally north of 60° N. While the North Polar ice cap thaws in summer and the land warms, allowing plant growth to a high latitude, the ice cap in Antarctica depresses summer temperatures, creating tundra conditions at lower latitudes, to 50° S in some areas. Temperatures on land fringing the Arctic rise well above freezing in the summer, supporting more than 400 kinds of flowering plants. Most of the ice-free zone of Antarctica is too dry and cold for vegetation, and life is limited to small patches of ice-free ground. Only two species of flowering plants are known; vegetation is primarily cryptogams. The Arctic has a variety of mammal and bird life. The Antarctic has several sea birds, but no land mammals. Permafrost is characteristic of the Arctic but is rare in the Antarctic because there is little moisture. The lack of trees in the Antarctic renders the concept of treeline irrelevant.

## Physical Environment

### Permafrost

Permafrost is ground that is permanently frozen. Widespread in the Arctic, permafrost covers half of both Canada and Russia. Beneath the boreal forest south of the tundra, permafrost is usually discontinuous or sporadic, depending on local conditions and insulating effects of vegetation. Poleward of treeline, permafrost is generally continuous, underlying the entire landscape with the exception of large bodies of water. The thickness of permafrost, which can extend down more than 2,000 ft (600 m), depends on several factors, including mean annual temperature, type of soil and rock, proximity to ocean, and topography. Permafrost can be wet—in which case abundant water forms ice bodies—or dry—in which case water in gravel, soil, or solid rock is limited. In either case, because of the cold climate, too little heat is available during the summer to completely thaw the soil.

Summer heat only thaws the surface part of the soil, below which the ground remains permanently frozen. The thawed surface, called the active layer, is often waterlogged because the permafrost below impedes drainage. Depth of the active layer varies but is generally 8–24 in (20–60 cm), except along rivers where it can be 6 ft (2 m) or more. On even slight slopes, the active layer may slowly flow, only 0.5–2 in (1–5 cm) per year, in a process called solifluction. Permafrost and solifluction exert both a negative and positive influence on tundra plants and animals. While continual disruption of surface soil creates difficulties for small seedlings, differences in topography or texture provide a variety of microhabitats.

Permafrost creates specific features on the landscape collectively called patterned ground, which can be circles, polygons, nets, hummocks, steps, or stripes (see Figure 2.1). Circles, nets, and polygons form on level ground. On steeper slopes, the patterns may be elongated until they become stripes. Although the particles making up the patterned ground may be sorted or unsorted, resulting in either a border of stones or no border at all, most patterns are caused by differential freezing of the active layer. Because they hold more water, fine-textured soils contract and expand with temperature changes more than coarse-textured soils do. As the soil freezes, any intermixed larger particles are pushed either upward or outward from a center. This process is most important in sorted patterns. Desiccation cracking, similar to how mud shrinks as it dries and cracks into polygons, is more significant in creating unsorted patterns. On a smaller scale, needle ice, which disrupts soil surfaces, occurs on moist surface soils when night temperatures drop below freezing. Several explanations for formation of patterns have been proposed, and different processes may produce similar results.

Usually small (15 in [35 cm] in diameter), circles may be outlined by coarse stones or by plants at the more stable edges. The centers where frost-heaving disturbs the soil are usually bare. Polygons can be various sizes. Desiccation cracking in summer creates polygons only 8 in (20 cm) in diameter, but polygons larger than 3.3 ft (1 m) are common and usually have ice wedges at their borders. Fine-textured soil between polygons may shrink and crack from either dryness or cold. The cracks collect water that when frozen enlarges the crack, exaggerating the polygon shape. Depending on the processes involved, either the edges or centers can be lower, and sometimes a lower center contains a pond in the summer. Frost-heaving may push rocks to the surface in the middle of the polygon. Larger rocks slowly move downslope away from higher centers and collect at the edges, creating a sorted polygon. Large polygons may be up to 100 ft (30 m) in diameter, with rims 3.3 ft (1 m) high. Different soil textures and water availability support different types of vegetation. Deeper snow accumulates in the troughs between polygons, providing more insulation and shelter for lemmings. Interconnected circles or polygons form a net or mesh pattern on the landscape.

Solifluction may create steps in the landscape where the active layer overrides more stable soil. Vegetation often concentrates at the base of the riser, which may be 15 in (40 cm) high, where snow accumulates and provides protection in winter. The tread, which is more exposed to wind, may be bare or may support cushion plants.

Other major features on tundra landscapes where water is abundant are hills called pingos. A horizontal subsurface layer of water freezes into a lens of ice. As it draws more water from the surrounding soil, the lens expands and pushes the overlying soil upward, forming a hill with an ice core. Although most are smaller, pingos may be

## Alaska Pipeline

Permafrost in Alaska posed serious problems for construction of the pipeline from the northern coast, where oil is extracted, to the southern ports from which it is shipped. The pipeline was elevated in areas of ice-rich permafrost, because a buried pipeline of warm oil would thaw frozen ground. Heat conducted down support columns of an elevated system, however, would also melt the permafrost, dislodge the columns, and disrupt the pipeline. The problem was solved by putting ammonia inside the columns. Evaporation and condensation processes involve an exchange of heat energy. Like water, liquid ammonia takes heat from its surroundings to evaporate, and the environment becomes cooler. When the ammonia gas condenses back into liquid, it releases that heat back into the environment. Inside the support column tube, liquid ammonia trickles down to well below ground level, where it absorbs heat from the ground and evaporates. The ammonia gas then rises inside the tube to above ground level, where the gas is chilled by cold arctic air surrounding the column. As the gas condenses back into liquid ammonia, it releases the heat taken from the ground into the atmosphere. The liquid then trickles back down to rewarm and reevaporate. In this never-ending cycle, the ammonia takes excess heat that would thaw the permafrost layer and transfers it to the air above. Ammonia has a very low freezing point and remains liquid in spite of temperatures that maintain permafrost. This closed system has successfully kept the permafrost frozen and the columns and pipeline in place.

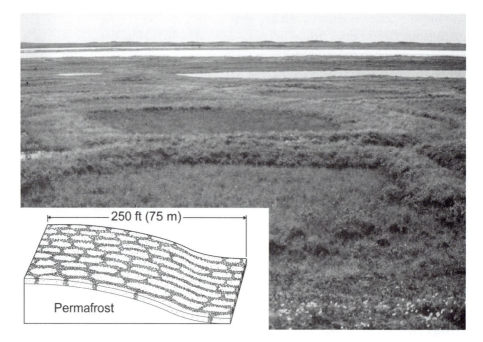

**Figure 2.1** Frost action in permafrost contributes to the development of patterned ground, from polygons to stripes, which determines vegetation patterns. *(Illustration by Jeff Dixon. Adapted from Sharp 1938. Photo courtesy of the U.S. Fish and Wildlife Service.)*

as much as 150 ft (45 m) high and 1,800 ft (550 m) in diameter at the base. Tops are relatively warm and well drained and are home to such animals as arctic ground squirrels and arctic foxes. A breach in the surface soil may let heat penetrate and melt the ice lens, resulting in a small hill with a pond at its center. More common in discontinuous permafrost and smaller than a pingo, a palsa is a low hill in peat or boggy areas that has an ice core.

## Climate

Temperatures in the Arctic and Antarctic are primarily controlled by the light regime. Both arctic and antarctic tundra regions experience 24 hours of daylight during some part of their respective summers, but the number of days with that extreme varies with latitude, ranging from one day at the Arctic and Antarctic Circles (66½° N and S) to six months at the poles (90° N and S). At 70° N and S, the sun is above the horizon continuously for two-and-a-half months and at 75° N and S for four months. The same figures apply to winter darkness, that is, days with no sunlight at all. Because of the angle of the sun's path, however, long hours of twilight in winter when the sun is only a few degrees below the horizon are characteristic. With no incoming solar radiation during the dark winter, temperatures drop. Even the continuous daylight of summer provides little warmth.

Temperature contrasts between summer and winter may be more related to degree of continentality than they are to latitude: Higher latitude areas in coastal locations, especially those affected by warm currents, are warmer in winter than lower-latitude regions farther inland. Mean temperatures of the warmest summer month vary considerably, but generally average between 40° F (4.5° C) and 55° F (13° C). Even though temperatures on a "hot" summer day occasionally may reach 70° F (21° C), the average for the short summer season is below 50° F (10° C).

For much of the year, the region experiences a negative energy balance (see sidebar Arctic Energy Budget). The low angle of the sun in the sky imparts little energy to the arctic landscape. The high albedo of ice and snow means that much incoming solar radiation is reflected and sent back to space. Infrared energy continues to radiate from both land and sea surfaces all year, even during winter darkness. The result is a continual energy loss most of the year. The energy balance does not become positive until after snow melts, usually in June, even though incoming solar radiation increases from the time of the spring equinox. This makes the summer growing season even shorter because about one-half of the summer's solar radiation comes before snow melts and is lost either by reflection off snow cover or is used as energy to melt snow rather than warm the air and plants. After snowmelt, albedo decreases abruptly and photosynthesis takes place continuously during the 24 hours of sunlight.

**Arctic Energy Budget**

Temperature of Earth's atmosphere is determined by the energy budget, basically a comparison of how much energy enters Earth's atmospheric system and how much energy leaves. In spring and summer when the sun is high in the sky and days become long, more energy reaches Earth than the amount that exits as infrared radiation. The surplus of energy causes temperatures to rise. In fall and winter when the sun drops lower in the sky and days become shorter, less energy comes in. Infrared energy is always being radiated back to space, however, and when the outgoing amount exceeds what comes in, there is a deficit of energy and temperatures begin to drop. Reflection, evaporation, and melting of snow complicate this pattern. Incoming solar radiation striking a white snowy surface will reflect back into space and not enter Earth's energy system at all. This phenomenon is called albedo. A sunburn on the bottom of your chin is from solar radiation reflected off snow or water. Incoming radiation may also be used to melt snow or evaporate water, and thus cannot be used to warm the air. That is why it is cooler around a swimming pool in the summer.

Cold winter temperatures in arctic tundra rarely have a direct effect on plant and animal life because most organisms are adapted to extremes of cold, dormant during the winter, or protected under an insulating cover of snow. Some animals migrate to escape extreme cold or lack of food. Because the tundra environment is on the fringes of the continents, winter temperatures are not as extreme as they are in lower-latitude boreal forest climates where many animals and birds live and hibernate, including some that migrate from the tundra. Because boreal forest trees lack a protective snow cover and are exposed to the weather, they must be more cold tolerant than tundra plants. The most significant factors affecting life in the Arctic are cool summer temperatures and the short time period during which temperatures remain above freezing. The short summer and lack of heat for active growth also affect chemical activity and food sources, indirectly limiting life as well.

Arctic precipitation is generally low because cold air cannot hold much water vapor. Coastal regions along warm currents, such as Iceland, Norway, and southern Alaska, are wetter, with up to 30 in (750 mm) or more precipitation a year, but the cold airmasses more common to the Arctic produce 10 in (250 mm) or less. Availability of moisture for plants, however, also depends on substrate and wind conditions. Wind often redistributes sparse snow cover, leaving some regions quite dry. Sublimation of frozen water into dry air also robs soil of needed moisture by eliminating snowmelt. Regions underlain by extensive permafrost, however, are boggy because water that cannot percolate into the ground remains in the surface soil.

## Circumpolar Climate Variation

*Temperature.* Although the climate is cold, temperature and precipitation vary according to latitude, airmasses, wind direction, coastal or continental location, mountains or valleys, and protected inlets or exposed coasts (see Figure 2.2). Because the sun angle is low, even small topographic irregularities cause differences in microclimate, such as sunny or shady slopes. The following transect describes changing climate conditions eastward around the Arctic region. Iceland and the northeast Atlantic all the way north to Svalbard and east to Murmansk, Russia, are influenced by the warm North Atlantic Drift current and are mild for the latitude. Eastward to the Ural Mountains in Russia, continental influence increases. Winters are colder, and the summer frost-free period is shorter. East of the Ural Mountains in northwestern Siberia, winters become even colder. Temperatures do not rise above freezing until June and the frost-free period is only 30–40 days. Northeastern Siberia between the Taymry Peninsula and the Kolyma River has the most extreme climate. Cold air drains from adjacent mountains, settling into valleys and causing severe temperature inversions with cold air underlying warm air. Summers are also cool.

East of the Kolyma River in northeastern Siberia and into Alaska and the Yukon, the landscape is characterized by many mountains and intervening valleys. The mountains are high enough to effectively separate cold Arctic airmasses in the north from warmer Pacific airmasses to the south. Northern coasts have an unmodified arctic climate, inner valleys have a continental influence, and Pacific coastal areas are moderate. In spite of many lakes in the area, the region between the Mackenzie River and Hudson Bay in the Northwest Territories and western Nunavut is continental with severe winters. Almost the entire Canadian archipelago, from Banks Island on the west to Ellesmere Island on the east, is surrounded by thick sea ice from December to April. The area is persistently cold but is not extreme. The eastern part of the Canadian Arctic, particularly the Ungava Peninsula in Quebec and Baffin Island, has more maritime influence because the sea there is not frozen. Cyclonic

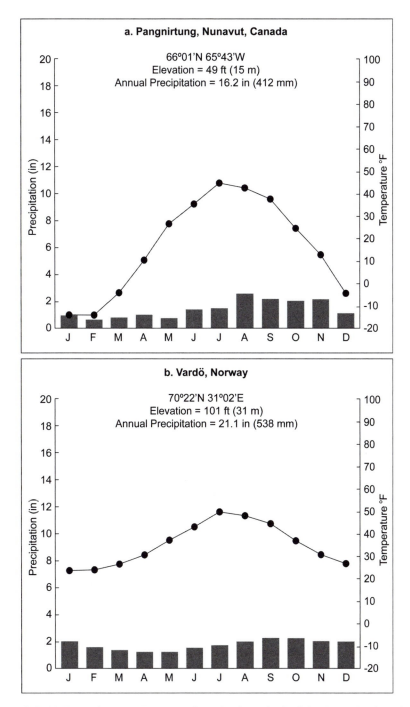

**Figure 2.2** (a) Pangnirtung, Nunavut, Canada, is typical of the Low Arctic, whereas (b) moderate temperatures and more precipitation at Vardö, Norway, are characteristic of maritime Arctic. *(Illustration by Jeff Dixon.)*

storms coming from the south bring warmer airmasses and temperatures near or above freezing even in winter, especially in the Ungava plateau.

*Precipitation.* Precipitation varies according to continental or coastal position and location with respect to wind direction and mountains, but is generally low. Most precipitation falls as snow, even in summer. Precipitation in Iceland, the northeast Atlantic, and Russia west of the Urals is high for the Arctic, 15–30 in (380–750 mm), and decreases toward the east. In northwestern Siberia east of the Ural Mountains, which are a precipitation barrier, annual precipitation drops to about 10 in (250 mm). The cold air continues to be dry eastward to the Kolyma River. Although winter air is extremely dry, ice fog over cities is frequent in December and January. Moisture generated by urban areas quickly freezes into the cold air. Precipitation in northeastern Siberia and adjacent Alaska varies because of topography. The north coast, influenced by Arctic airmasses, is fairly dry. Interior mountains have upslope precipitation, and south-facing mountains have extensive snowfall. Anchorage, Alaska, gets 5 ft (1.5 m) of winter snow, the equivalent of 6 in (150 mm) of rain.

Because of the dominance of high pressure and absence of cyclonic storms, little precipitation falls in northwestern Canada and the Canadian islands during the winter months, although frequently blowing snow reduces visibility. Cyclonic storms do bring up to 2 in (50 mm) of rain in the short summer, and increased moisture from melting snow and ice causes frequent fog. Higher rainfall in the eastern Canadian Arctic is caused by mountainous topography and paths of cyclonic storms from the Great Lakes or Atlantic coast. Precipitation primarily falls in summer and is variable, depending on storm tracks.

Strength and occurrence of wind depends on topography and frequency of cyclonic storms. It is not exceptionally windy in the Arctic, but wind is more noticeable because low vegetation does not block it. Infrequently a warm winter wind will melt snow, which then recrystallizes into a hard crust, preventing animals like muskoxen and lemmings from accessing food. Except for steep slopes, wind-exposed ridges, and among the tallest shrubs along rivers, most of the Arctic is snow-covered from late August to May or June, or even early July.

## Tundra Soils

Soil development is slow and limited because of little plant cover, a short growing season, permafrost, and the short time land has been free of glacial ice. Soils are thin, young, and nutrient poor. The cold and short growing season inhibits both chemical weathering of mineral material and biologic activity. Sparse vegetation contributes little organic matter and nutrients. Clear soil profiles rarely develop because frost-heaving and solifluction continually overturn soil layers. Soil classification systems for the Arctic are varied but generally mirror patterns of vegetation. Minimally developed spodosols (podzols) exist on well-drained soils with a deep active layer dominated by dwarf-shrub heath and dwarf birch. Inceptisols (Arctic brown soils) occur on uplands and dry ridges where cushions and heath shrubs grow. Other inceptisols (tundra soils), where the dominant process is gleization, are common where drainage is impeded. They support cottongrass, tussock grasses, dwarf-shrub heath, and some sedges. Histosols (bog and half-bog soils), which also develop through gleization, occur along shores of lakes and ponds where drainage is better and peat accumulates from sedges and mosses. On drier and better-drained areas, a soil profile begins to develop, with an iron and clay layer at the surface. In polar deserts, calcium carbonate and magnesium frequently accumulate on undersides of rocks as the surface dries in summer, especially in areas of sedimentary rocks and recently uplifted marine sediments. In contrast to most tundra soils, which are acidic, these polar desert soils are basic, mainly because they lack nitrogen and phosphorus.

## Tundra Vegetation

Arctic tundra plant life is dominated by dwarf shrubs, mosses, lichens, sedges, and perennial forbs. The region was still covered by ice only 8,000–15,000 years ago. The circumpolar Arctic is divided into two major zones according to latitude, climate, and vegetation. Covering the largest area in lower latitudes, the Low Arctic has the most variety of growthforms. The active layer is deep on rolling upland, allowing good drainage and root penetration. Vegetation cover is 80–100%, with less on rocky continental shield areas. The general appearance of the landscape is that of a grassland with cottongrass tussocks and birch and willow shrubs up to 5 ft (1.5 m) tall (see Figure 2.3). Subshrubs, mostly heaths, are shorter, less than 8 in

**Figure 2.3** The tundra near Churchill, Manitoba, in Canada is grassy with stunted trees and low shrubs. *(Photo by author.)*

(20 cm) high. Fruticose lichens such as reindeer moss and snow lichen grow in dry sites and mosses such as haircap grow where it is wet.

While Low Arctic vegetation is commonly called tundra, the higher latitudes of the High Arctic are often described as polar semidesert and polar desert. The High Arctic is mostly bare rock, angular from frost action; plants are herbaceous. Most of the area is polar semidesert, covered by lichens and mosses. Flowering rosettes, cushions, and mat-forming forbs, mostly saxifrages, are only found in protected sites and are no more than 3 in (8 cm) high. In true polar deserts, even cryptogams are sparse, and total cover is 0–3%.

On a large scale, the circumpolar tundra zone has similar plant communities; however, regional and local differences occur.

## Tundra Animals

### Large Mammals

***Caribou and reindeer.*** North American caribou and European reindeer, the same species, are the most common large mammal (see Figure 2.4). They spend winters at the forest edge and summers on the tundra. Bulls average 350 lb (160 kg); females are smaller. Unlike other deer, both sexes have antlers, and except for pregnant females, both genders shed them in fall. Pregnant females keep their antlers until late spring when they calve, using them as deterrent to keep others away from foraging areas scraped free of snow. Calving grounds are on the tundra, which is often still snow-covered in May when births occur. Calves are precocious and can run soon after being born. After a week of nursing their mothers' rich milk, they can forage on their own.

Large, concave hoofs give support in deep snow or marshy land, and a thick coat with hollow hairs provides insulation. The caribou winter diet consists of starchy lichens rich in carbohydrates, which gives them energy to survive the cold. During spring and summer, they eat high-protein shoots of sedges, grasses, dwarf willows, and birches. Mouth muscles sort living from dead plant tissue; indigestible parts are spit out. To avoid overgrazing one area, the animals continually migrate in large herds of thousands, trampling the landscape along regular routes up to 500 miles (800 km) long. Wild herds in the tens of thousands still exist in America, where their spectacular migrations create conflict between conservation and development. Most European reindeer were domesticated 3,000 years ago, although a few wild animals still remain in Fennoscandia and Russia.

***Muskoxen.*** A fixture in the high north for thousands of years, muskoxen once shared the tundra with mastodons and mammoths (see Plate IV). They are big animals, up to 900 lb (408 kg) and 7 ft (2 m) long, but with short legs. Both sexes have horns. Glands between the bulls' eyes produce the musky odor for which they are named. Long brown or black hair almost touching the ground combined with a soft, wooly undercoat provide superb insulation. Lying down, the animal does not even lose enough heat to melt the snow and ice beneath it. Muskoxen live on the tundra all year, eating a variety of grasses and shrubs. In summer, herds consist of about 10 animals; they increase to 15–20 in winter. One male dominates each herd, breeding with the females, which each bear one calf in spring. The calf matures fast, being able to graze within a week, but continues to take its mother's milk for a year. The animals' natural method of defense, backing into a circle with horns extended outward, successfully protects calves in the center from arctic wolves, but made them easy prey for rifle-wielding humans. Populations once on the verge of extinction have been brought back through conservation efforts. Unlike caribou, they do not undertake long migrations, but spend winter on higher ground where winds blow vegetation free of snow.

***Brown bears.*** Brown bears, grizzly, Kodiak, and Kamchatka bears are all the same species regardless of geographic location, although scientists recognize several subspecies. A mature male can weigh 900 lb (400 kg) and stand 9 ft (3 m) tall on its hind feet, but females are smaller. Found in both forest and tundra habitats throughout the Northern Hemisphere, they may be brown, cream-colored, or black. Brown bears are distinguished from their close relative the black bear by their dished face and a prominent hump on their shoulders. Long, curved claws and muscles in the hump are adaptations to digging out roots, insects in decaying wood, or small mammal burrows. Best known for gorging on migrating salmon in Alaska, brown bears are not picky eaters. Although classified as carnivores, they are omnivorous, eating plants, fruit, insects, birds, rodents, moose, caribou, and carrion, a varied taste that lures them to human garbage dumps in populated areas. Sedges, grasses, and roots are tasty treats on the tundra. Because of their size and food requirements, they are solitary except for mothers and cubs. Although not true hibernators, they lower their body temperature and metabolism in winter and go into torpor.

## Caribou Parasites

Caribou have few natural enemies. Golden Eagles and bears prey on calves, while packs of wolves can overpower adults. When temperatures rise to 50° F (10° C), swarms of summer insects in clouds of millions plague the animals and interfere with feeding. Experiments have shown that mosquitoes can extract up to 4 oz (115 g) of blood from one animal in a single day. They try to escape by twitching, running, occupying windy locations, clustering together, or immersing themselves in snow or water. The two worst insects plaguing caribou are warble flies and nose bot flies. Warble flies lay eggs on the caribou's leg hairs. After burrowing into the skin, larvae travel beneath the surface to the animal's back, where they cut air holes through the skin to the surface. They grow to about an inch long beneath the animal's hide. When development finishes, the larvae pop through the skin and pupate on the tundra into adult flies. Female nose bot flies lay their eggs in the caribou's nostrils, from which hatched larvae move to the throat, to take advantage of the warm, moist environment found there throughout the winter. In March, the larvae grow quickly and are eventually coughed out by the caribou to pupate into adults on the tundra.

**Figure 2.4** Reindeer, shown here in Lapland in northern Norway, and caribou in North America are the same species. *(Photo by author.)*

***Polar bears.*** As tall as 10 ft (3 m) and weighing up to 1,700 lb (770 kg), adult male polar bears are the largest members of the bear family. Females are smaller. Population estimates are between 22,000 and 27,000. Their only enemies are humans and now quite possibly climate change. Polar bears are at the top of the arctic food chain, spending winter hunting seals on pack ice by waiting at breathing holes or stalking them as they rest. They prefer ringed seals, but also eat bearded seals, and if hungry, will eat reindeer, birds, and beached whales. A bear can eat 100 lbs (45 kg) of seal blubber at a single sitting, and if hunting is good, leave the rest of the meat for scavengers. Preference for fat has an advantage in that its digestion releases metabolic water, important in an environment where water is either salty or frozen.

Polar bears are well adapted to frigid arctic conditions. An insulating layer of blubber up to 4.5 in (12 cm) thick helps maintain normal body

**Polar Bear Colors**

Although usually appearing to be some shade of white, polar bear fur has no pigment and takes on the hue of its surroundings, slightly yellow-orange from the low-horizon sun or bluish in light filtered through clouds or fog. In late summer or fall, bears can appear gray because oils from their prey have stained the coat. Bears in zoos may have a green tint because of algae growing in the fur. Polar bear skin is actually black, and if fur is damaged, black patches may show through.

temperature of 98.6° F (37° C), the same as a human's. Under extreme cold or windy conditions, animals curl up in a snowbank for additional protection, covering their uninsulated muzzle with a furry paw. Their bulk, with small ears and tail, helps to conserve heat. Two different layers of fur provide two layers of insulation. The undercoat is dense fur, and longer guard hairs are hollow. With all this insulation, bears can become too warm, especially when running, and shed excess fur for summer. Feet are covered with small bumps, and sometimes hair, giving them traction on icy surfaces. Black claws are sharply curved for both clinging to ice and catching seals.

Because of harsh conditions, reproduction is slow. An adult female may have only five litters in her 18-year life span. Although they can create a den on drifting pack ice, pregnant females usually dig snow dens on coastal land in fall. They give birth during the winter, not eating, drinking, or defecating until spring when the new family leaves the den. Normally, the litter is two cubs. Like all bear cubs, they are small, 12 in (30 cm) and 1 lb (0.5 kg), blind, helpless, and totally dependent on mother's milk, which has 31% fat content. By April, they can weigh 50 lb (22 kg). Cubs remain with their mother for two-and-a-half years while they learn to hunt and survive. Neither males nor females hibernate.

Their circumpolar distribution is limited to ocean areas frozen into sea ice most of the year and adjacent land in North America and Eurasia. Because they hunt seals, they prefer shallow water near the shore or the edge of the ice pack that does not freeze too deeply. Although classified as marine mammals because of preferred habitat on sea ice, most live on the tundra in summer while others remain on ice floes all year. Polar bears need sea ice on which to hunt, and without it they have no access to their prey, seals. As such, they are vulnerable to climatic warming.

## Small Mammals

Small rodents are plentiful. Many remain active in winter but need an insulating layer of snow. Lemmings and voles survive under the snow, where they are hunted by weasels. The same areas that have an insulating snow cover, however, are most subject to spring flooding. Lemmings that fail to move their nests to higher ground

### Polar Bear Jail

Churchill, Manitoba, is near the site where waters of Hudson Bay freeze first because of the fresh water outlet from the Churchill River. Bears migrate from other parts of Canada to get on the ice first near Churchill. They have fasted all summer, living off accumulated fat, and by early to mid-November, they are hungry and ready for seals. Although dozens of bears may congregate near Churchill, the population can disappear overnight when ice conditions are right. Bears initially gathered at the town dump where garbage was available, but they began to cause trouble. A Polar Bear Alert program now uses noise primarily to scare bears away from town. If a bear continues to be a problem, especially if a mother is teaching her cubs bad habits, the animal is humanely trapped and taken to polar bear jail, a hangar-like facility away from town. When the Bay freezes, they are released. If the facility is full, helicopters airlift the bears in mesh cargo nets and deposit them in remote locations, with the hope that they will not return. Adults are sedated to keep them calm, and their eyes are coated with petroleum jelly for protection from cold, dry air during the airlift. Small cubs too delicate for the net are transported inside the helicopter.

in spring risk drowning. The tundra vole, which can swim, lives in the wettest areas. Brown lemmings occupy better-drained polygon ridges, and collared lemmings live on well-drained rocky uplands. Narrow-skulled voles are found on the best-drained sandy soils along streams. Several voles live in shrubby heath, willow, or alder thickets. No animals especially suited for digging, such as gophers, inhabit the arctic tundra because permafrost, extremely cold winters, and shallow snow cover leave no unfrozen layer through which to dig. Arctic hares in Low Arctic tundra spend the winter eating twigs that extend above the snow cover.

*Lemming cycles.* Food chains in the tundra are short and simple. Lemming population cycles that peak and crash every three to six years are a good example of the delicate balance in tundra ecosystems. Small mouse-like animals, lemmings are only 5 in (13 cm) from nose to short tail. Although they have a reputation of committing mass suicide, they are merely migrating in an attempt to escape overcrowding. The animals do not hibernate, but remain active beneath snow all winter eating green shoots of sedges and grasses. They also mate all year, and within two years, a female can produce up to 14 litters of eight pups each. However, due to predation and drowning, most lemmings do not live that long. Nonetheless, because pups double their birth weight in just four days and females can become pregnant while still nursing, a colony can potentially increase its population 100 times in a single winter. Lemmings must eat twice their weight in food every day, which has consequences for vegetation and underlying soil. Digestion is inefficient, and 70% of the intake is deposited as nitrogen-rich waste, which nourishes tundra vegetation. By cutting dead vegetation to gain access to new, green shoots, lemmings aid the decay process. As a result, lemming foraging areas can become lush, but if too much plant cover is removed, thawing of permafrost leads to denudation.

Lemming population increases lead to more predators. Populations of arctic fox, weasels, Snowy Owls, and Pomerine Jaegers increase. Wolves, which usually attack larger prey, will also eat lemmings because they are so abundant. Owls catch lemmings on the surface, while the slim bodies of weasels allow them to pursue lemmings in their tunnels. Eventually the lemmings must migrate to find food or die. Crowding, insufficient food, or less-nutritious food inhibit breeding, and the lemming population crashes. A corresponding decrease in predator populations follows. During the bust period, vegetation recovers and permafrost is reestablished. Lemmings in Fennoscandia migrate in the thousands, while elsewhere they move in smaller groups.

### Birds
Many birds, including waterfowl, jaegers, and ptarmigan, are found in the Arctic, and some, such as Snow Geese and ptarmigan, influence the vegetation by their grazing activities. Millions of geese, swans, ducks, gulls, and waders live in the tundra in summer. Waterfowl are the first to arrive in spring, followed by shorebirds and land birds. Birds that nest in the tundra include several species of plover,

sandpiper, geese, and ducks. By necessity, birds must nest on the ground, vulnerable to scavengers and predatory mammals. Several birds of prey such as falcons, hawks, and Snowy Owls share the same nesting areas with prey species. Food includes insects, fish, and aquatic vegetation. Mosquitoes, craneflies, springtails, and stoneflies are plentiful because the numerous ponds and bogs are good breeding sites. Many tundra birds change their diets according to season and availability. They feed on both seeds and insects in spring, concentrate on insects in summer, and change to seeds in late summer. Food intake may be determined by habitat, or habitat may be selected according to food preferences. Willow Ptarmigans eat willow twigs and birch buds and leaves, while the Rock Ptarmigan's diet is alder catkins and buds. Geese eat grass, sometimes the whole plant, leaving the land bare except for mosses and lichens. With less grass, cyanobacteria, an important nitrogen-fixer, grows on the surface. Because bird excrement is rich in nitrogen, surviving or colonizing vegetation grows well on nesting grounds.

Among the very few passerines, Lapland Longspur, Snow Bunting, wagtails, and pipits breed on North American and Eurasia tundra. Many birds fly long distances. Undertaking an exceptional 21,000 mi (33,800 km) migration each year, the Arctic Tern breeds in arctic tundra but winters in subantarctic seas. Peregrine Falcon, a raptor, also summers in the tundra. Arctic birds have various ways of caring for young. Either males or females may sit on the nest and feed the chicks while the other flies away. Some produce two clutches, one each for the male and female to tend, in the hopes that some might survive. Leaving the breeding grounds before the end of the summer season may be a way to reduce demands on scarce food resources.

Although many birds nest on the tundra, few stay all year. Gyrfalcon, the world's largest falcon, eats arctic hare and ptarmigan, while ravens survive on carrion, droppings, and human garbage. Snowy Owls, which nest and lay eggs before snow melts in spring, prefer lemmings and other small animals. Ptarmigan congregate on bare land or areas where the snow is thin because they cannot scratch through deep snow for food.

In spite of continuous light or darkness during arctic summers or winters, both birds and mammals maintain a daily activity cycle that corresponds to a 24-hour period. Daily activities in small mammals are controlled by an innate circadian rhythm rather than cues from changing light levels.

## Animal Effects on Vegetation

Activities of mammals and birds can alter plant communities and extensively damage tundra vegetation. Concentrated grazing by reindeer has the biggest impact on reindeer lichen. Because lichens grow slowly, 0.2 in (6 mm) per year, where winter grazing is intense, grazing can totally destroy the carpet in three to four years. It may require five years of other growth, mostly grasses, before lichens begin to regenerate, and 15–20 years for total restoration. However, most wild reindeer herds are self-regulating. They graze lichens only under extreme cold and maximum snow cover, quickly moving to summer pastures when conditions allow. Grazing by domesticated herds does much more damage because of their concentration in limited areas.

Arctic fox has a local but strong influence on vegetation. The animal's deep entry passage into its burrow disturbs permafrost, resulting in a deeper active layer. Aeration and accumulation of excrement enrich the soil, which then supports a dense cover of several grasses and forbs. Bean Geese and White-fronted Geese prefer cottongrasses and other sedges, only eating the succulent parts next to the root. The accumulation of organic matter in stems and bird droppings then influences vegetation growth through the addition of nutrients. Geese have large feeding areas and usually do not do much damage. If concentrated, however, geese can devastate an area, leaving it barren, trampled, and covered in droppings. Birds of prey use the same spot for years, and effects of altered vegetation beneath roosts are quite visible in different and taller growths of forbs.

Lemming population fluctuations are large and significant. When lemming populations peak every three to four years, they overgraze and reduce the vegetative cover. They normally do not eat the less-nutritious mosses, but still damage them in winter under the snow cover by gnawing through the moss layer to reach the succulent parts of grasses and forbs. Since lemmings can destroy local areas while leaving frozen parts of others intact, they contribute to the hummocky surfaces so common in arctic tundra. Over time, the untouched turf becomes thicker with shoots, forming a hummock. The less nutritious upper stems that lemmings cut and discard become dried plant litter, an important source of organic matter in the soil.

## Human Influences

Air pollution, including radioactive fallout, accumulates in the Arctic due to wind patterns and lack of precipitation to clean the air. Lichens absorb moisture and nutrients directly from the air, and because of slow growth, accumulate toxic materials over many years. Radioactive particles introduced into the food chain when lichen is eaten by reindeer or caribou can be transferred to human populations eating the meat. Mining, exploration, roads, and other development disrupt the fragile balance of tundra ecosystems. In spite of that fragility, alteration and destruction due to human activities is surprisingly limited due to sparse populations. Introduced plants rarely last more than one or two seasons because they are not adapted to the climate or soils. Vehicle traffic compresses and eventually destroys the mat of vegetation, and the main dominants of moss, lichens, dwarf shrubs, and shrubs are usually replaced by grasses and forbs. In such altered areas, the number of plant species is only one-half to one-third that of natural tundra. Alteration of the permafrost by buildings, roads, or change in vegetation cover can be significant. As the insulating layer is destroyed, more ice melts, resulting in a phenomenon called thermokarst, in which the surface soil collapses into voids vacated by ice cores. Problems are only intense near the few populated areas in the Arctic.

Global warming is visibly affecting tundra environments. Ponds in the tundra on Ellesmere Island have been drying up over the last 25 years, and salt content in the water has increased due to accelerated evaporation. Sea ice and permafrost in several arctic or antarctic environments are melting at unprecedented rates. Coastal regions left unprotected by disappearing sea ice are now exposed to erosion, and seawater encroachments on formerly fresh water lakes threaten wildlife habitat. Temperature changes in the oceans will affect land. Snow appears to be melting earlier in the summer in some places, and permafrost temperatures are rising.

## Regional Expressions of Arctic Tundra

### North America

Tundra occupies about 20% of the North American continent, from 55° N along Hudson Bay to Alert on Ellesmere Island at 82° 03′ N. Primarily in Canada, it stretches from 61° W in Newfoundland to 168° W in Alaska (see Figure 2.5). In this large territory, climate, topography, ice cover, soils, ecosystems, and diversity of plants and animals all display much variation. The most expansive region is Low Arctic, the gently sloped uplands that have good drainage and deeper soils to permafrost. High Arctic tundra is found on the northern Canadian islands, including Ellesmere, Baffin, and Queen Elizabeth.

*Climate.* In winter, cold temperatures over the continent create a high pressure, which keeps moist airmasses away. With summer warming, the high pressure

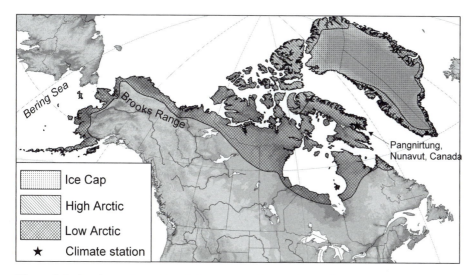

**Figure 2.5** Arctic tundra in North America and Greenland. *(Map by Bernd Kuennecke.)*

weakens, and the Arctic Front—the southern limit of cold, dry air—roughly coincides with treeline. Cyclonic storms and moist airmasses can then enter the Arctic and bring summer rain. This seasonal cycle of pressure and storms results in dry winters and wetter summers. The northwestern Canadian islands, however, are dominated by cold, dry arctic air all year and are surrounded by sea ice. The southeastern Canadian Arctic has a warmer maritime climate with more cyclonic storms. The Canadian mainland is continental, with a greater seasonal temperature range and much warmer summers than experienced farther north.

The Brooks Range falls on the border between arctic and boreal forest biomes, vegetatively defining north and south Alaska. Cyclonic storms are frequent on the south coast, where the Bering Sea receives high precipitation, but infrequent on the Arctic coast because mountains block the storms. North Alaska experiences cold temperatures with deep temperature inversions. Ice fog is a hazard in industrial areas when temperatures drop below 5° F (−15° C). The west coast facing the Bering Sea is less severe.

In eastern Canada, precipitation is limited except for maritime Baffin Island, but because most precipitation falls as light rain or mist every few days in summer, soils rarely dry out except in the High Arctic. Soil moisture is usually high, especially where permafrost impedes drainage.

Differences in biotic communities between Low Arctic and High Arctic can be attributed to differences in severity of climate (see Table 2.1). Radiation, mean annual temperature, summer and winter temperatures, length of growing season corresponding to time of snowmelt, and precipitation all decrease from south to north. Even with a longer period of summer daylight at higher latitudes, the lower sun angle and high reflectivity off snowy surfaces diminishes radiation gain farther north, resulting in lower temperatures. Mean annual temperatures in the Low Arctic, both in maritime locations and the more continental interiors, are higher than those in the High Arctic. Both summer and winter temperatures are slightly warmer in the Low Arctic. Low Arctic receives more precipitation, and snow cover is generally deeper. Because snow melts sooner, the growing season is one to two months longer in the Low Arctic. The date of snowmelt is significant in

**Table 2.1 North American Arctic Tundra Climate Summary**

| Climate Characteristic | Low Arctic | High Arctic |
| --- | --- | --- |
| Mean Annual Temperature | 20° F (−6.5° C) | 10° F (−12° C) |
| January Mean Temperature | −10° F (−24° C) | −22° F (−30° C) |
| July Mean Temperature | 50° F (10° C) | 40° F (4.5° C) |
| Annual Precipitation | 18 in (450 mm) | 6 in (150 mm) |
| Depth of Snow Cover | 14 in (35 cm) | 6 in (15 cm) |
| Growing Season | 3–4 months | 1.5–2.5 months |
| Date of Snow Melt | mid-June | late July or August |
| Depth of Active Layer | 8–80 in (20–200 cm) | 8–60 in (20–150 cm) |

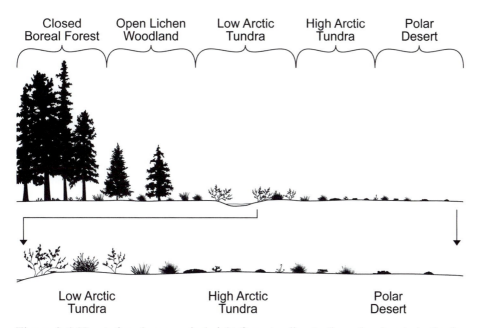

**Figure 2.6** Vegetation decreases in height from treeline to the polar deserts in the far North. *(Illustration by Jeff Dixon.)*

determining the growing season. Average depth of the active layer, however, is similar in both subdivisions.

Vegetation varies according to temperature, aridity, and snow depth, usually becoming shorter and more sparse with increasing latitude (see Figure 2.6).

*Plant communities of the Low Arctic.* North of the closed boreal forest, an open lichen woodland of widely spaced trees marks the beginning of treeline, here a broad zone more than 100 mi (160 km) wide. Fruticose lichens grow 6 in (15 cm) thick on treeless ground. Still farther north, the woodland thins to small, scattered groves in sheltered sites where snow accumulates. Black spruce dominates these treeline groves in most of Canada but is replaced by white spruce in the Yukon and Alaska. Low Arctic tundra vegetation fills the nonforested areas between groves of trees. Although fires are necessary to maintain the open woodland of the treeline zone, they kill shrubs, mosses, and lichens. Shrubs resprout within 2–10 years, but it takes 60–200 years for the lichen community to regenerate.

With the exception of Barrow on the Arctic coast, the Low Arctic includes all of Alaska, mainland Canada, southern Baffin Island, and southern Greenland. The term tundra refers to Low Arctic vegetation with a complete cover of mainly woody species, sedges, and grasses, and excludes the cushion-plant vegetation that is more typical of High Arctic and polar deserts. The landscape is grassy with low or dwarf shrubs. Shrubs are significant except in wetlands, where they are replaced by sedges, bryophytes, and grasses.

Scientists recognize several major vegetation types in which shrub height corre-lates with snow depth. Tall shrub tundra consisting of alders, birch, and willows about 10 ft (3 m) high, with a dense understory of forbs and grasses, is most com-mon along river bottoms, near lakes, and other protected areas where warmer soil contains more nutrients and the active layer is deeper. Species vary slightly accord-ing to geographic location and subspecies are recognized. The understory can be nodding hairgrass and matgrass. This vegetation type is important habitat for moose in northwest Canada and Alaska. Arctic hare and ptarmigan also feed on the willows.

Low shrub tundra with various combinations of dwarf birch, low willows, heath, forbs, and graminoids is found primarily on rolling uplands in Alaska and northwestern Canada beyond the forest tundra zone where snowcover is not as deep as in protected areas. An open canopy of shrubs averaging 20 in (50 cm) high is dominated by bog birch and blue willow. Ground cover is made up of clumps of carex sedge and cottongrass tussocks, and tiny heath subshrubs only 6 in (15 cm) high (see Plate V). Arctic bell heather is found where snow is late to melt. Mosses and fruticose lichens complete the ground cover. In northeastern Canada, limited snow cover and abrasive winds support different species and a more limited plant assemblage.

In areas with minor snow cover, dwarf-shrub heath tundra may or may not include cottongrass tussocks. Small heath shrubs grow on medium-drained soils. Species vary with geographic location and include mountain azalea and Lapland rosebay. Evergreen leaves on several species are diagnostic of this community. Graminoids include red fescue, highland rush, and arctic bluegrass. Large areas of western and northern Alaska, especially the Mackenzie River delta, are covered with a community variant that includes cottongrass. Cottongrass is conspicuous because of its white tufts, but heaths still dominate. A well-developed cryptogam layer of mosses, including several *Sphagnum* species, is present. Especially when they flower in early summer, cottongrass communities are important grazing ground for caribou, providing good nutrition for lactating females.

Regions with low relief, poor drainage, and a thin active layer over permafrost, such as the coastal plain and outer Mackenzie Delta support graminoid-moss tun-dra (mires) with plants 12 in (30 cm) tall. Because of shallow permafrost, this region is characterized by patterned ground, polygons, nets, and stripes. These wet-land meadows are dominated by water sedge, few-flowered sedge, and cottongrass, but they also include marsh cinquefoil. Different soil textures support different plants. The fine-grained soil in the centers of the polygons or in cracks between them are too saturated for much plant life other than some mosses. Lichens, dwarf shrubs, and forbs grow on the coarser, raised, and better-drained rims.

Wind-swept slopes and ridges with little to no snow cover are a cushion plant–cryptogam polar semidesert, with low cushions or mats. This community, similar to the High Arctic and limited in the west, is more common in the east where more bare rock and gravel occur. Mountain avens plants, less than 2 in (5 cm) high but more than 3 ft (1 m) in diameter, can account for 80–90% of the vascular plant cover, especially on calcareous soils. Associated low-stature perennial forbs are northern anemone, viviparous bistort, and moss campion. Lichens are common.

***Plant communities of the High Arctic.*** High Arctic refers to the high-latitude Canadian archipelago and the northern tip of Greenland. Because of more arid conditions and less protective snow, plant cover is sparse. Several vegetation types occur, two of which are small areas of tundra vegetation similar to those communities found in the Low Arctic but with few to no tall shrubs. Cushion plant–cryptogam and cryptogam-herb semidesert cover most large areas of the High Arctic. Herb barrens constitute small areas of true polar deserts.

Graminoid-moss tundra, found wherever drainage is impeded, such as river terraces and coastal lowlands, forms major grazing grounds for muskoxen. Adjacent lakes and ponds provide breeding sites for waterfowl and shorebirds. Sedges are the most important plants, but mountain foxtail and Fisher's dupontia grasses are also common. Saxifrage cushions, cottongrass, and lousewort are widespread, but mountain avens and prostrate arctic willows are restricted to hummocks with better drainage. Mosses and cyanobacteria are abundant, but lichens are rare. Mires in the southern and eastern High Arctic are dominated by sedges, but in the northwest Canadian islands, they are dominated by dupontia.

Dwarf-shrub heath tundra has fewer heath species than in the Low Arctic. Heath tundra needs snowbed sites that melt by early July, situated near rocks that absorb and provide summer heat, situations rarely found in the High Arctic. Arctic bell heather is most characteristic, but other species include short-leaf sedge, arctic willow, purple saxifrage, and mountain avens cushions. Mosses are plentiful. Many fruticose lichens, including several *Cetraria* species, arctic finger, and whiteworm are members of this community. Heath tundra in northern and central Baffin Island has a richer flora than elsewhere but is still dominated by arctic bell heather.

Polar semideserts are not barren like true polar deserts, but have some plant cover, a richer flora, and more soil development. They cover about one-half of the southern Canadian islands and one-quarter of the northern islands. Cushion plant–cryptogam vegetation is similar to that found in limited sites in the Low Arctic. Predominantly growing on warm gravelly surfaces with a deep active layer, it becomes a minor component where rocky surfaces, fine soils, cold climate, or late snowmelt occurs. Vegetation is dominated by mats or rounded cushions of arctic mountain avens, with associated cushion plants of several saxifrages, sandworts, and starworts. Vascular plants, however, are less important than cryptogams. Lichens and mosses are numerous and varied. Semideserts provide the main habitat for Peary's caribou, collared lemming, ptarmigan, and numerous passerine birds.

Cryptogam-herb semidesert is the general vegetation type on low rolling uplands and beaches with finer soils rather than gravel or rocks. Vascular plants account for only 5–20% of the total; cryptogams make up 50–80%. Vegetation is dominated by bryophytes and both crustose and foliose lichens, in contrast with the dominance of fruticose forms in the Low Arctic. Yet, the general appearance is that of a grassland because graminoids of mountain foxtail and northern woodrush are visually dominant. Upland sedges are totally absent. Forbs include rosettes or cushions of mouse-ear, draba, sandwort, rooted poppy, and saxifrage species. Bryophytes are abundant, including several mosses. A black cryptogamic crust—a mixture of black crustose lichens and cyanobacteria capable of fixing nitrogen—is common on the soil surface. In the High Arctic, seedlings of vascular plants do best in deep mosses and desiccation cracks where conditions are more moist and there is less needle ice. More seedlings die in summer due to soil drought than due to winter cold.

Found largely in the Queen Elizabeth Islands, polar deserts in the High Arctic are dominated by herb barrens that form a mosaic with semidesert vegetation. About 98% of the area is bare soil and rocks, justifying the term desert. Herb barrens are determined by the lack of soil development, low soil-nutrient content, needle ice, dry surface, and a short growing season of just one to one-and-a-half months. The few vascular plants are found in relatively protected areas of desiccation cracks or near stones in patterned ground. Vascular plants may include draba and rooted poppy rosettes and red sandwort and purple saxifrage cushions. Cryptogams cover less than 1% of the ground, a major contrast with semideserts where cryptogams dominate. The barrens were probably still covered by ice during the Little Ice Age 130–430 years ago. Not even lichens grow yet on the rocks.

***Animals.*** North American mammals are those typical of circumpolar tundra. Caribou is the major large herbivore in the Low Arctic, joined by muskoxen at higher latitudes. Muskox graze sedge-moss meadows, and in winter, paw through as much as 20 in (50 cm) of snow to feed. Small herbivores in the Low Arctic include arctic hare, both brown lemming and collared lemming, and arctic ground squirrel. Brown lemmings graze on graminoids and mosses in wet areas, while collared lemmings eat forbs and dwarf shrubs in drier areas. The only mouse-like species in the High Arctic is the collared lemming, most populous in raised-center polygons with deep snow in the troughs. Because of harsher conditions, this species has fewer litters per year than its relatives in the Low Arctic. Their primary foods are mountain avens, herbaceous willow, and purple saxifrage. Arctic hares are more common in the Low Arctic, where they feed on willows.

Predators in the Low Arctic include brown bears that venture into the tundra to feed primarily on roots and grasses. Smaller animals that prey on rodents such as lemmings include the least weasel and short-tailed weasel, both of which can maneuver into narrow burrows. Because lemming populations are lower and nesting birds with eggs are less common, fewer carnivores inhabit the High Arctic, in terms of both species and total numbers. Least weasels are not found in the High Arctic, but short-tailed weasels are. Arctic fox is common to both environments, as is an occasional timber (or arctic) wolf. By hunting in packs, wolves are capable of

**Figure 2.7** Snow Goose with nest in the High Arctic. *(Courtesy of the U.S. Fish and Wildlife Service.)*

taking larger animals such as caribou. Foxes search out small mammals and follow polar bears onto sea ice to scavenge leftover seal kills. Avian predators outnumber mammalian predators. Low Arctic predatory birds numerous in high lemming years include Snowy Owls, jaegers, Glaucous Gulls, and Short-eared Owls. These birds winter elsewhere, the Snowy Owls being the first to arrive in early spring, the rest returning to the tundra by the time snow melts. All but the Short-eared Owl are found in the High Arctic. The only predators staying in the tundra all winter are the least weasel, short-tailed weasel, and arctic fox. Polar bears only use the tundra during summer.

Ptarmigan are birds common to both environments. Willow Ptarmigan is more common in the Low Arctic where willow shrubs are a significant food source while Rock Ptarmigan is the major species in the High Arctic. Snow Geese breed in the High Arctic but winter along the western coast of the United States (see Figure 2.7). Other birds found in the Low Arctic are also typical of circumpolar tundra.

### Greenland

Covering more than 23 degrees of latitude from 59° 46′ N to 83° 04′ N, the northern most land on Earth, Greenland has considerable climate variation. A distinct north-south gradient in temperature and precipitation develops, especially in winter. Almost 80% of Greenland is ice cap, actually islands buried in ice. Areas closest to the ice cap are more continental in climate, while coastal regions are varied due to a cold current on the east and south coasts and a warm current off the southern third of the west coast. Summers are colder on the east coast.

This large continent in the path of westerly winds splits storm tracks. Radiation cooling from the ice cap causes a persistent high pressure in the north, while the southern part experiences cyclonic storms that move north along the west coast before crossing over the ice cap. Directly in the path of cyclonic storms generated between polar and maritime airmasses, the southwest coast is wet, with 30 in (760 mm) or more of precipitation annually. Coastal climate is also influenced by whether or not land is adjacent to sea ice. All coastal stations have moderate temperatures compared with the rest of the Arctic. Nord (81° 36′ N) is the coldest with a mean annual temperature of 2.5° F (−16.4° C). Stations on the south coast have mean annual temperatures around freezing. Inner fjords are warmer than the coasts in summer, due both to warming effects of downslope (föhn) winds and less oceanic influence. Although these strong winds originate on the ice cap, they increase in temperature as they descend and bring warm, dry conditions, raising temperatures as much as 35° F (20° C) in a few minutes. Such increases in winter can melt snow, and expose vegetation where snow cover is light. Where accumulations are deeper, only the surface melts, and then it subsequently refreezes forming an ice crust. The interior gets less precipitation, has a shorter season of snow cover, and a more continental temperature regime, that is, higher temperatures in summer and lower in winter. Warmer summers and a longer growing season are important for plants there.

Northern Greenland is polar desert, with a low mean annual temperature, strong winds, low precipitation, and high evaporation.

Greenland's topography and history has affected its flora and fauna. Almost all of the current ice-free land was glaciated during the Pleistocene and was exposed only in the last 10,000 years. The ice cap is a barrier to most organisms, limiting colonization and distribution areas. The ice-free edge of the continent is dissected by fjords and glaciers into peninsulas and islands, further limiting migration. Because of this relative isolation, fewer species of both plants and animals are found in Greenland than in other parts of the Arctic. Caribou, arctic hare, and arctic fox are the only animals found throughout ice-free Greenland. Arctic wolf, short-tailed weasel, collared lemming, and muskox are present only in High Arctic Greenland.

Most mammals originated in Eurasia and came across the Bering Strait to North America, and then migrated to Greenland. Three migration routes are reflected in plant and animal distributions. The route from North America via Ellesmere Island over Naras Strait allowed access primarily for High Arctic species. Davis Strait from Baffin Island and Labrador brought in Low Arctic and boreal species from North America. Other low Arctic and boreal species migrated from Europe via North Atlantic Islands. Wintering grounds of migrating birds indicate their probable origins. Wheatears winter in tropical Africa, and most likely colonized Greenland from the Old World. Snow Buntings fly from northeastern Greenland to the southern Russian steppes and to the Great Lakes, indicating that they probably colonized from both the east and west. Greenland Redpolls winter in both North America and Europe, also indicating a double colonization source. Lapland Buntings, which winter in North America, probably also colonized from there.

Any species that migrated to Greenland from Europe must have been maritime or quite adaptable. Several species that are primarily continental in Eurasia have become maritime in Greenland, indicating a westward migration. Elements of Greenland's flora that have come from the west, from continental North America, have remained continental in character. The south and southeast coasts of Greenland have a majority of European elements in the flora, both because of proximity and similarity of climate with northwest Europe. The western flora is more similar to that of the North American continent.

Tundra vegetation in Greenland varies according to topography and climate. Most of the ice-free area is mountainous, with rolling uplands and deep glacial troughs, creating local climates and habitats. Greenland has no extensive landscapes comparable to those found in the rest of the Arctic. Geology also varies, and substrate may be many types of rocks or sediment. Soils are podzolics, arctic brown, and polar desert, with low organic matter. Because of aridity, some soils are salty, and salt crusts may be caused by excessive summer evaporation. Wet soils supporting sedges, grasses, and mosses are rare and only found in broad, flat valleys in west-central and eastern Greenland.

Inland heads of fjords, which are more continental and have July mean temperatures over 50° F (10° C), support low shrub tundra of downy birch and Greenland ash. Some plants only occur around the hot springs common in basalt areas of west-central and eastern Greenland, well north of their normal climate limits. A willow scrub and a variety of herbaceous plants are restricted to this habitat. Hot spring habitats may have served as refugia for tundra plants during colder climate periods.

Most plants and animals have narrow environmental requirements. For example, wading birds need both snow-free nesting grounds and abundant insects for food. The only area with good conditions for both is north of Scoresby Sund. Farther north, little food is available, and farther south, both too much snow and cool summers limit insect numbers. Therefore, most waders in Greenland occur in the High Arctic rather than Low Arctic or subarctic. Ptarmigan in eastern Greenland migrate according to availability of food. They spend the summer breeding season in the interior near the fjords, where they feed on viviparous bistort, especially the protein-rich bulbils. Throughout the short summer, they follow the bistort bloom into the mountains. In winter, they move to open plateaus and plains on the outer coast where vegetation is blown free of snow and the major food is mountain avens, arctic willow, purple saxifrage, and bistort bulbils, according to snow cover. Caribou in western Greenland move from coastal wintering grounds to inland summer calving grounds just as high-quality graminoids emerge.

Three areas of Greenland have been well studied, and two others provide important habitat. The southern tip of Greenland, considered subarctic, is shrubby.

Birch about 5 ft (1.5 m) tall can be found in sheltered snow-covered places. However, no seedlings are produced and plants propagate vegetatively. The underlayer consists of dwarf shrubs, grasses, or forbs. Most of this area and habitat was exploited or cleared for agriculture and pasture as early as 1,000 years ago.

On the southwest coast at about 66° N, the area centered on Søndre Strømfjord is Low Arctic with a pronounced continental-maritime gradient. The western part has high mountains, up to 4,500 ft (1,400 m), with alpine glaciers, putting the eastern low hills and U-shaped valleys in rainshadow. A steep gradient of precipitation, temperature, and duration of snow according to topography is reflected in vegetation, from grassy meadows and dwarf shrub on the coast to grasslands with salt pans and lakes in the drier interior. Inner fjords that are warmer in summer support low shrub communities dominated by willow and birch, with an open canopy up to 2 ft (0.5 m) high. This general region is the most important territory for caribou in Greenland. Their grazing has changed blue willow scrub into bluegrass communities. The caribou population is unstable due to overgrazing and the lack of their natural predators, wolves.

Jameson Land in eastern Greenland north of Scoresby Sund at about 70° N is a transition from Low to High Arctic. The gentle plateau dipping west is a large lowland area with steep, moist slopes and inland valleys with a warmer and drier continental climate. Up to 1,300 ft (400 m), it is dominated by dry dwarf-shrub heath with a 20–75% cover of vascular plants. Dominant species include arctic bell heather, bog bilberry, and mountain azalea (see Plate VI). Moist sites also have mosses. Snowbed vegetation within 6 mi (10 km) of the coast supports Bigelow's sedge, bistort, and arctic willow. The drier east has open dwarf-shrub heath. Even drier sites in the east are fellfield with alpine bearberry, mountain avens, arctic willow, and moss campion. Jameson Land is one of the most important areas for muskox in the world. Winter foraging grounds have little snow cover, and summer sites are thick forb communities where deep winter snows impart summer moisture. This region also provides important July molting areas for Pink-footed Geese and Barnacle Geese. The birds are separated by habitat—Pink-footed Geese along the coast and rivers with an open view and Barnacle Geese on smaller rivers and lakes surrounded by hills—and feeding zones. They both eat sedges but in different parts of the meadows.

Disko is an island off the west-central coast, about 70° N. It is a high (up to 5,900 ft [1,800 m]) basalt plateau dissected by U-shaped valleys and steep slopes. The west has a High Arctic maritime environment, cool and foggy, with solifluction stripes and frost boils on the surface. Vegetation is an open, mossy heath dominated by arctic willow and arctic bell heather along with herbaceous plants. South-facing slopes have dense forb vegetation and blue willow scrub, especially around hot springs where water temperature can reach 50° F (10° C). Species more normally found farther south grow well in these habitats. Peary Land at the northern tip of the continent at 83° N also has a High Arctic environment. Precipitation near the coast, which always falls as snow, is less than 8 in (200 mm), decreasing to less than 1 in (25 mm) in the interior. What little snow falls is wind blown, exposing much bare land. The drier interior has mountain avens, heath, and graminoids. Snowdrift sites, significant because they provide the only water in summer, support sparse communities with sedges, cottongrass, arctic bell heather, and mosses. Cover in both is less than 5%. On the Arctic coast, which is more humid with frequent fog, mosses and lichens grow in patterned ground cracks.

## Eurasia

*Fennoscandia.* Although little permafrost occurs in Fennoscandia (the peninsula that includes Norway, Sweden, and Finland), mean annual temperature below freezing and summer months below 50° F (10° C) qualify the area as a tundra environment. Much of the mountain backbone of the peninsula is high enough for alpine tundra, and arctic and alpine zones merge in the north (see Figure 2.8). Fennoscandia is positioned in storm tracks between warm Atlantic air and cold polar air. Maritime conditions prevail on the coast, while the lee side of the mountains in the interior is drier and more continental. For example, Hardangervidda, the high-elevation plateau in southern Norway, has 60 in (1,500 mm) of precipitation on the west and 20 in (500 mm) on the east. The coast receives autumn precipitation caused by cyclonic storms, while summer convectional showers bring moisture to the interior. The lowest precipitation occurs in the northeast, with only 12 in (300 mm) received annually at Kautokeino, Norway. Except in coastal locations, winter temperatures may drop below −40° F (−40° C). The length of the growing season is 120 days or less, and on the Hardangervidda, it is a mere 74 days. Most areas have more than 200 days with snow cover, but depth and coverage, from inches to several feet, varies with wind. The time it takes for snow to melt affects the length of the growing season, and some snow may remain all summer.

### Iceland

At 65° N, Iceland is on the southern boundary of the Arctic. While the island is basically a lava plateau, plant distributions are complicated by fjords, isolated mountain peaks, lava flows, glaciation, wind-blown sand deposits, and settlement since 900 AD. For centuries the highlands have been used for summer grazing of freely roaming sheep. A small landmass affected by a warm current, Iceland has mild temperatures for the latitude. July averages 45° F (7° C), and January means drop to only 19° F (−7° C). Annual precipitation is high for the Arctic, 25–50 in (600–1,200 mm), and snow is deep, averaging 40 in (100 cm). Because of glaciation and volcanic activity, soils are gently rolling till-covered moraines, glaciofluvial deposits, or lava flows. Plant communities are similar to those found on mainland arctic environments, but differences exist. Basalt limits acidity, so little or no sphagnum moss and few heaths grow on recent lava flows. Lichens are poorly represented in heath vegetation, partly due to grazing by reindeer. Widespread on the island, Arctic fox is the only native land mammal. The fox has two color phases; one becomes white in winter while the other remains bluish white all year. Mink, introduced for fur farming in the 1930s, have escaped and are currently found in the vegetated highlands but not on the barren plateau. Reindeer have been introduced, and an estimated 4,000 live in the highlands. There are no native small mammals. A field mouse was introduced and its feral distribution today is not known. The bird fauna is rich and varied. Approximately 80,000 Pink-footed Geese breed in large colonies on the plateau, and 10,000 Whooper Swans breed along lakes and ponds.

**Coal Mining in the Arctic**

Halfway between Norway and the North Pole, Svalbard was an international whaling base in the seventeenth and eighteenth centuries. At 74° 81′ N, you would expect the island group to be bitter cold and uninhabited, but because it is in the path of the warm North Atlantic Drift, temperatures are relatively mild. The warmest summer month averages 40° F (4.5° C), while winters average 7° F (−14° C). Plants are typical of circumpolar Arctic. The islands have a wealth of natural resources, especially coal but also wildlife. Although it has been part of Norway since 1920, several nations have operated mines, but only Norwegian and Russian companies still extract coal. More than half of the population (total population 1,701 in 2005) is employed in coal mining and related service industries. About 60% of the land is covered by glaciers and snowfields, and ice floes often block entrances to ports. There are no muskoxen, hares, wolves, owls, falcons, or ravens. The single rodent, a mouse that occupies bird cliffs, was possibly introduced. There are also few invertebrates, notably no biting flies that harass reindeer. The single mammalian predator is arctic fox, which preys on birds and eats carrion. Two endemic subspecies, Svalbard reindeer and Svalbard Ptarmigan, are of conservation significance. The reindeer probably colonized from northeastern Greenland and the Canadian archipelago and has been isolated for 40,000 years. Even though few mammals live on Svalbard, visitors can see some of the largest concentrations of birds in the North Atlantic. Migrating birds that nest on the tundra include Pink-footed Goose, Barnacle Goose, and Brant Goose. Polar bears can be seen along the coast in summer. About 65% of the islands is environmentally protected in nature reserves, national parks, and bird sanctuaries.

Well-defined vegetation belts characterize Fennoscandia tundra. The subalpine includes the upper or northern limit of boreal forest, a treeline with downy birch, which is also typical of arctic tundra in Russia, Iceland, and Greenland. Treeline varies from 4,000 ft (1,200 m) in southern Norway (61° N) to sea level in the north. The crooked stems of birches are less than 30 ft (10 m) high, and the zone is the equivalent to krummholz in North American alpine areas. In more continental areas, the treeline and krummholz tree is Norway spruce. Bog birch, common juniper, and several heath species such as Scotch heather and blueberries, make up the shrub layer. An herb layer consists of nodding hairgrass, matgrass, and carex sedges. Mosses are found in wetter areas, replaced by lichens where it is drier.

The low alpine is similar to the subalpine but without trees. With lower summer temperatures, shrub size is much smaller, 10 ft (3 m) high down to 8 in (20 cm). Thickets have tall forbs such as monkshood, buttercup, globeflower, and graminoids. The ground layer, unless covered by litter or shaded, consists of mosses and liverworts. The most common heath association in the low alpine is dominated by European blueberry with a ground layer of bryophytes. Because less organic matter accumulates at higher elevations, the low-alpine belt is the upper limit of mires. Mire vegetation includes bog rosemary, bog birch, cottongrass, small cranberry, and sphagnum moss. Graminoids dominate moderate snowbeds, but extremely long-lasting snowbeds may have mosses, the colorful mountain lichen, moss plant, and buttercups. More continental areas have a fruticose lichen ground cover of Iceland moss and *Cladonia* lichens.

Many communities have an upper limit between the low- and mid-alpine belts (see Figure 2.9). The mid-alpine is dominated by meadows of highland rush, but Bigelow's sedge, sheep fescue, and other graminoids also occur, with minor representation of heaths. This belt is absent in

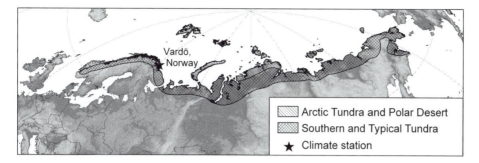

**Figure 2.8** Except for the mountains in Fennoscandia, Arctic tundra in Eurasia is limited to a narrow coastal zone. *(Map by Bernd Kuennecke.)*

Finland. An abrupt transition from low-alpine to mid-alpine exists, noted by the absence of several plant communities, but a gradual transition separates mid-alpine and high-alpine belts in which many of the same species are present. Plants growing in both mid-alpine and high-alpine belts include moss campion and saxifrage cushions and spiked trisetum and carex graminoids, along with alpine sorrel and several buttercups. Dwarf willows are the only shrubs in the high-alpine zone where the number of plant species decreases with elevation. Plant cover, dominated by lichens and mosses, is incomplete. Solifluction, polygons, and stripes

**Figure 2.9** Vegetation on high mountains in Norway, here at Gudvangen, merges with arctic tundra in northern Fennoscandia. *(Photo by author.)*

make the soil rocky and unstable but still able to support a wide variety of vascular plants.

A more circumpolar flora exists on peaks, while more strictly Eurasian species appear at lower elevations. Few plants are endemic to the region. Plants originating in the western Arctic probably survived the Pleistocene by taking refuge on the northern and southern mountains that extended above the ice. The central mountains were too low and covered by ice. Some species are found in only one high-alpine area, indicating the location of their refugia.

Animal life in Fennoscandia is more varied than in most of the Arctic, reflecting the connections to alpine environments. However, not all species are present in both the arctic and alpine regions, and mammals are few. Wild reindeer have been extinct in Sweden since 1860 and in Finland since 1900. Today, they are only found in small Hardangervidda areas in Norway. However, domesticated herds are many and widespread. Small mammals include Norway lemmings, famous for their massive migrations, shrews, and voles. Digging animals are more common than in the rest of the Arctic due to the lack of permafrost. The root vole is found in moist areas near streams or mires in the low- and mid-alpine zones. Other voles occupy other habitats. Mountain hare are found all the way to the high-alpine zone, and like ptarmigan, will burrow into snowbanks for warmth.

Several native mammalian predators occur, primarily in alpine environments, but most large species are endangered. Wolverines were hunted almost to extinction but are now protected, with populations primarily on the mountainous frontier between Sweden and Finland. Wolves are nearly extinct, and only a few brown bears still live in Sweden's forests. Lynx is the most common predator in the subalpine and low-alpine zones. Lynx, brown bear, and wolf are mainly forest animals, occasionally venturing into the tundra. Smaller predatory mammals more commonly found in both arctic and alpine tundra are ermine and other weasels, red fox, and arctic fox, although few arctic fox remain. All predatory mammal populations fluctuate with prey populations.

Nesting birds are plentiful, their local presence determined by vegetation and habitat. In wet areas, waterfowl and shorebirds dominate. Whooper Swans and Bean Geese are found in wet areas of birch forests, while Mallards prefer smaller lakes surrounded by willows. Willow Ptarmigan are found in willow thickets, while Rock Ptarmigan prefer rocky slopes. Waders such as Golden Plover and Red-necked Phalarope congregate around bogs and adjacent heathlands. Passerines such as Snow Bunting, Meadow Pipit, and wheatears frequent several habitats.

Because many predatory birds only reproduce when prey is plentiful, their numbers correlate with populations of small mammals. Rough-legged Buzzards and Short-eared Owls catch rodents, while Golden Eagle and Gyrfalcon both prey on ptarmigan. Lemmings are prey to Snowy Owl and Long-tailed Skua in the Arctic. Ravens are opportunistic, catching live prey and eating carrion. Many predatory birds are now protected by law.

European common frog is the only amphibian found above treeline, but it is rare even in the low-alpine zone. Reptiles are scarce. The common viviparous lizard lives in the low alpine, but grass snake is limited to the subalpine zone.

*Russia.* Latitudinal belts of tundra are limited to a relatively narrow northern strip on the Eurasian continent north of the Arctic Circle. Terminology is confusing because all authors do not use the same terms to define Russian Arctic zones. Three categories of tundra (southern, typical, and arctic), as well as polar desert to the north and forest tundra to the south, are described here. All zones are not represented equally across the continent.

Temperature in all zones is strongly influenced by degree of continentality (see Table 2.2). Mean annual temperatures are similar, as are January means, which are considerably warmer than the more continental boreal forest farther south. Tundra zones thus reflect differences in July mean temperatures, which are succeedingly cooler from south to north. July in the forest tundra averages 54° F (12° C), above the 50° F (10° C) limit for tundra, which explains the presence of trees. Both the western and eastern parts of the Russian Arctic are slightly warmer due to warmer oceans nearby, particularly the North Atlantic Drift and Barents Sea on the west. Summer begins when mean air temperatures rise above freezing, which occurs in early June in southern tundra but is delayed to early July in polar desert. Temperatures drop below freezing in September. The frost-free season is short, ranging from two months or less in polar desert to three-and-a-half months in southern and forest tundra. The western Russian Arctic has a longer frost-free season, four-and-a-half to five months from mid-May until mid-October, due to the influence of the ice-free North Atlantic Ocean.

**Table 2.2  Russian Arctic Tundra Climate Summary**

| CLIMATE CHARACTERISTIC | SOUTHERN TUNDRA | TYPICAL TUNDRA | ARCTIC TUNDRA | POLAR DESERT |
|---|---|---|---|---|
| Mean Annual Temperature | 10° F (−12° C) | 10° F (−12° C) | 10° F (−12° C) | 10° F (−12° C) |
| January Mean Temperature | −15° F (−26° C) | −15° F (−26° C) | −15° F (−26° C) | −15° F (−26° C) |
| July Mean Temperature | 48° F (9° C) | 48° F (9° C) | 39° F (4° C) | 34° F (1° C) |
| Annual Precipitation | 13 in (330 mm) | 13 in (330 mm) | 9 in (230 mm) | 7 in (175 mm) |
| Depth of Snow Cover | 16 in (40 cm) | 16 in (40 cm) | 9 in (23 cm) | 9 in (23 cm) |
| Depth of Active Layer | 33 in (85 cm) | 30 in (75 cm) | 21 in (55 cm) | 16 in (40 cm) |

Precipitation varies with latitude, longitude, topography, and distance from the ocean. In general, polar deserts are driest and tundra zones are wetter. Precipitation is highest on the mountains in the central Taymyr peninsula and on the Barents Sea Coast, where up to 20 in (500 mm) may be received annually. About one-third falls as rain in July and August. Relative humidity is high, typically 80–90%, because cold air quickly reaches saturation.

Snow cover in tundra zones lasts 200–280 days, remaining the longest amount of time in colder Siberia and shortest amount of time in the warmer west. Because of generally light winter precipitation, snow cover is not deep, but it is variable depending on geographic location. Less snow falls in the north and polar desert, and more on the flatter southern tundra. European Russia has deeper snow (up to 24 in [60 cm]) because it is both warmer and wetter. Frequent strong winds redistribute the snow, blowing ridgetops free and piling snow in swales. Snowmelt begins in early May in the west, but not until the end of June on arctic islands. Some sheltered areas retain snow all summer. Summer snowfalls are light, and the temporary cooling is too brief to harm most plants.

The typical Siberian arctic landscape is a gently rolling plain with many streams, ponds, and lakes. All but the westernmost area is underlain by continuous permafrost, and various types of patterned ground are characteristic. Polygonal mires are a pattern of polygons up to 50 ft (15 m) in diameter with rims about 6.5 ft (2 m) wide and 3 ft (1 m) high. Deep cracks in the rims fill with water that freezes in winter, perpetuating the polygon, while depressed centers contain ponds or bogs in summer. The landscape is large scale. One mire may consist of 50–100 polygons. Frost boils are also common. Frost-heaving pushes bare soil to the surface, like a bubble in boiling water. The bare top may be 25 in (60 cm) in diameter, surrounded by vegetated soil 4 in (10 cm) higher. Boils are separated by narrow, shallow troughs. Whole landscapes can consist of frost boils. Groups of small pingos, up to 30 ft (10 m) wide and 1.5 ft (.5 m) high, are relics of past glaciation and create a hummocky landscape with continuous vegetation cover. Larger pingos are 600 ft (180 m) in diameter and 230 ft (70 m) high.

The depth of the active layer varies but generally decreases with latitude. The deepest active layer, 50 in (130 cm), is on south-facing slopes in southern and typical tundras. Peat bogs have the shallowest active layers, 20 in (50 cm) in the south, decreasing to only 8 in (20 cm) in the north.

Southern tundra is the ecotone between tundra and boreal forest, the northern limits of treeline, and is widest on the west and east where climates are slightly warmer. Tree growth is sparse, with krummholz and prostrate trees along rivers. Treeline species vary with geographic area but include downy birch, willow-like *Chosenia arbutifolia*, several species of larch, Siberian spruce, and Mongolian poplar. This zone is characterized by shrub communities with a three-tier stratification. Taller shrubs, 2 ft (0.6 m), are several species of birch and willow, as well as alder and mountain alder. Shrub height depends on depth of winter snow, and plants are taller where snow provides more protection. Many dwarf shrubs, 6 in (15 cm) tall,

such as arctic bell heather, mountain crowberry, and bog bilberry, grow under-
neath as a second layer. Mountain avens cushions, several carex sedges, and cottongrass are also common. A low ground cover, 4 in (10 cm), consists of a rich
variety of fruticose and foliose lichens, along with mosses and some liverworts.

In addition to the more typical shrub community, other plant communities
occur in other situations. Large, flat areas may have palsa bogs with a grassy cover
of cottongrasses and sedges and water-loving mosses such as *Calliergon* and *Drepanocladus.* Peat hills are formed by *Sphagnum* species. Warmer south-facing slopes
have meadows with a rich mix of grasses, legumes, and other forbs, both arctic and
boreal species. Forest enclaves on the European arctic islands or along rivers have
typical treeline species according to geographic area. Northeastern Russia has tussock communities of carex sedges and cottongrass.

Southern tundra has a diverse assemblage of animals. Birds are abundant, especially water or shore birds such as White-fronted Goose, Brant Goose, Stone
Plover, and sandpipers. Surface-feeding, or puddle ducks, are also common. Many
wading birds and passerines are insectivores, but Willow Ptarmigan feeds on willow stems and buds.

Typical mammals are voles and lemmings. Voles are widely distributed in
southern tundra, especially in European Russia, where their presence increases certain forbs and grasses. Predators are numerous compared with the relatively few
prey species. Arctic fox, Snowy Owl, Pomatorhine Skua, and Rough-legged Buzzard all prey on lemmings, and their populations are generally tied to the lemming
cycle. Skuas only breed when lemming populations are high. Foxes depend on a
variety of food, but increase in population when lemmings peak. Snowy Owls also
breed more in high-lemming years, and rarely nest in low-population years. The
relationship between lemmings and weasel populations is not known.

Proximity to several biogeographic provinces contributes to the rich variety of
species in the southern tundra, and many nonarctic elements are present in both
the flora and fauna, among them several boreal forest mosses and many birds and
insects, especially in the forest enclaves. Plants more typical of northern tundra
zones are absent or restricted to snowbeds. Circumpolar species predominate, but
more geographic variation exists here than in northern tundra zones. Compared
with European Russia, Siberia is more strictly arctic.

Typical tundra is also called moss-lichen tundra because of the prevalence of
mosses and lichens and its low stature, less than 8 in (20 cm) high (see Figure 2.10).
This zone, which lacks trees altogether, has its widest extent in the west; it is fragmented in the east. Closed shrub thickets are limited to river valleys or protected
areas. Vegetation is low willow shrubs with semiprostrate birches and willows.
Mosses are prominent and characteristic, making a 5–10 in (13–25 cm) thick layer
on the soil. The liverwort *Ptilidium ciliare* is a common codominant in the moss
cover, where many species combine to form a mosaic. The moss turf can either be a
continuous layer or be interrupted by frost boils with bare spots. Because the moss
layer is thick, vascular plants must have long rhizomes to reach through to the soil.

**Figure 2.10** Typical tundra in Russia is characterized by a thick layer of lichens, mosses, and liverworts. *(Photo by author.)*

Carex sedges are common, as are some dwarf shrubs, mainly Lapland bell heather, polar willow, and alpine avens cushions. The same fruticose and foliose lichens are found here as in southern tundra. Some dwarf heath shrubs grow in protected locations, but they rarely flower in this colder zone. A greater circumpolar component in the flora, similar in both species and appearance all across Arctic Russia, exists in this belt, in contrast to more geographic variation in southern tundra. In river valleys, there may be low willow thickets or low moss carpets of water-loving species, with a few flowering plants of alpine sorrel, buttercups, and saxifrage cushions. Drier fellfields have mountain avens cushions but no mosses, and warmer south-facing fellfield slopes support meadows with a variety of grasses and forbs.

Birds typical of southern tundra, such as White-fronted Goose and Stone Plover, are absent from typical tundra. Without shrubs, no birds or insects dependent on shrubs can survive. Small mammals are limited to lemmings. The moss is home to a rich variety of invertebrates, including springtails, mites, spiders, and rove beetles, but the sparse undergrowth supports few insects.

Arctic tundra is found on the northwestern mainland coast, and on Arctic islands only in the east. The absence of shrubs and even most dwarf shrubs simplifies the vertical structure of the community. The moss turf is only 1–2 in (3–5 cm) thick, and vegetative parts of flowering plants exist within the moss, not extending above it except for an occasional flower stem to 3 in (10 cm). No stratification exists, just one mixed layer of mosses and forbs. Sedges such as the cottongrasses common to typical tundra are replaced by grasses and forbs. Growthforms include tufts, cushions, and dense mats. The tiny dwarf shrub, polar willow, grows totally protected within the moss layer, with its leaves lying on top. Mountain avens are found only on fellfields. Fisher's dupontia grass gains dominance as carex sedge numbers decline. The liverwort disappears and is replaced by an abundance of *Drepanocladus uncinatus* moss. Except for sphagnum, other dominant mosses remain the same as in typical tundra. Fruticose lichens are replaced by crustose types. Ground may be 50% bare, and no communities have a closed vegetation cover. Plants find shelter in cracks caused by permafrost action, and polygon patterns are outlined by vegetation. Depressions may be bare if snow fails to melt early enough to allow plant growth. Few additional habitats occur in arctic tundra to support different vegetation. South-facing slopes have a flora much like the general community.

The most northern arctic environment, polar desert, has limited extent, occurring primarily on the north half of Novaya Zemyla and the western arctic islands, and on Cape Chelyuskun in Taymyr. This extreme environment is characterized by low summer temperature, little precipitation, a shallow active layer, and a short growing period. Biotic diversity is low, with the number of vascular plant species in local floras being about 50, one-half of that found in arctic tundra and one-quarter of that in typical tundra. Less diversity is seen in mosses and lichens. Up to 95% of the ground is bare; only the cracks bordering polygons are vegetated. Cryptogams dominate. Fruticose lichens include *Cetraria* species, snow lichen, and whiteworm lichen; *Parmelia* is an important foliose lichen. The moss layer is only 2 in (5 cm) thick and lacks stratification. Flowering plants are sparse, found either within the moss layer or lying flat on its surface. A uniform species composition prevails throughout the polar desert, the vegetation only changing in density depending on habitat. Only two or three land birds nest in the region. No earthworms, spiders, beetles, sawflies, or craneflies inhabit this tundra belt.

Reindeer are the most numerous large herbivore on the Eurasian tundra. In spite of sometimes heavy grazing, wild herds have little influence on tundra vegetation, and most winter pasture for both wild and domesticated reindeer is in or at the edge of the northern boreal forest. Reindeer change their diet during migration. On Taymyr and breeding grounds in the north, they feed on grasses and forbs; in the southern tundra, they eat sedges, including cottongrass, and willows. In boreal forest wintering grounds, they feed on lichens. Isolated herds on Novosibirskiye, Novaya Zemyla, and Sibiryakova islands subsist all winter on polar willow shoots and buds, because no shrub, lichen, or meadow communities occur there. The population uses multiple feeding grounds, so it rarely damages tundra vegetation. Trampling only occurs in concentrated trail areas like land bridges between lakes or river crossings.

## Antarctica

Antarctica is a large continent with two geologic zones (see Figure 2.11). East Antarctica is a stable continental shield composed of metamorphic and sedimentary

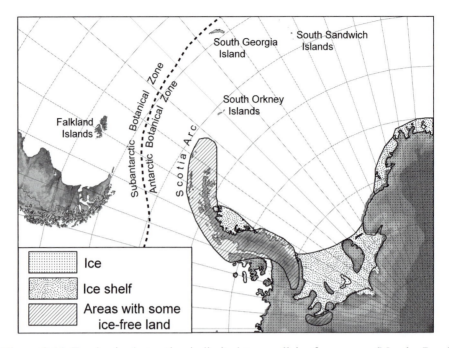

**Figure 2.11** Tundra in Antarctica is limited to small ice-free areas. *(Map by Bernd Kuennecke.)*

rocks, while West Antarctica is a mountain belt that includes the Antarctic Peninsula. Scotia Arc, a chain of peaks forming several islands off the northwestern part of the Peninsula, has a long history of volcanism. It extends north to the South Shetland Islands and northeast to the South Orkney Islands, South Sandwich Islands, and South Georgia, almost to South America. Less than 5% of Antarctica is ice-free. Most exposed land occurs along the coast, on islands, and on narrow coastal strips of the Antarctic Peninsula, where the ocean moderates temperatures, but dry valleys and nunataks (mountains above the ice) can also be found.

The northern boundary of the Antarctic tundra environment is best defined as the 50° F (10° C) February isotherm, which occurs at the latitude where cold ocean currents of the Southern Ocean meet warmer waters from the north, initiating a change in both plankton and types of sea birds. It is also the position of the southern Polar Front, where polar air meets mid-latitude air, roughly 55° to 60° S. These latitudes, excluding the tip of South America, make up the subantarctic zone and a transition from Southern Hemisphere cool-temperate climates to Antarctic conditions. The subantarctic and transition zones also include the Kerguelen Islands at 50° S in the Indian Ocean and South Georgia at 54° S in the South Atlantic Ocean. The Antarctic, subdivided into northern and southern zones (generally south of 60° S), includes the South Sandwich Islands at 58° S and Bouvetøya at 55° S in the South Atlantic Ocean. As in the Arctic, mean annual temperature is below freezing.

Diversity of terrestrial life is low compared with similar latitudes in the Arctic due to the small extent of ice-free area, severe climate, climate history, and geographic isolation. Organisms that disappeared during more extensive glacial cover have been unable to recolonize the continent. The milder subantarctic zone has higher plant diversity, with 24 graminoids, 32 forbs, 250 mosses, 150 liverworts, and more than 300 lichens. The northern Antarctic zone has only one graminoid, one forb, 75 mosses, 25 liverworts, and 150 lichens. Only two species of flowering plants, Antarctic pearlwort and Antarctic hairgrass, grow on the continent, as far south as 68° 12′ S. Sunny north-facing slopes are the only areas that support turf-forming mosses and vascular plants. Diversity in the southern zone is reduced to 30 mosses, one liverwort, and 125 lichens. No graminoids or forbs are found there. Heat, wind, moisture, substrate, slope, nitrogen, and disturbance by birds and mammals determine which plants actually grow at a particular site.

In the northern or coastal zone—which includes the northwestern Antarctic Peninsula, adjacent islands, some islands in the Scotia Arc, and Bouvetøya—only one summer month averages above freezing. The southern boundary of the zone is where the average temperature of the warmest summer month (January) fails to rise above freezing. The coastal zone is fairly wet, especially in the north; it receives at least 20 in (500 mm) of precipitation and regularly experiences relative humidity over 80%. Although soils are shallow, frost action, solifluction, and accumulations of peat are common. The west coast of the Antarctic Peninsula, where it is relatively moist, supports 100 kinds of lichens, while the drier east coast has many fewer. Two main communities exist in this predominantly maritime zone. A grass-moss combination that may also include lichens is dominated by the only two flowering plants in Antarctica, both limited to this zone. Sheltered north-facing slopes lower than 350 ft (100 m) elevation and some flat areas protected by winter snow support grass-moss or grass-lichen-moss communities. The several subdivisions of lichen-moss association are based on different species and different percentages of lichens versus mosses. Lichen crusts and fruticose lichens coexist with moss cushions. Moss communities include a haircap moss and *Chorisodontium aciphyllum* turf with peat accumulations up to 3.3 ft (1 m) thick, or thin moss carpets of *Drepanocladus*, *Brachythecium*, and *Acrocladium*. Wet areas may have a moss carpet of *Calliergidium*, *Calliergon*, and *Drepanocladus*. Fruticose lichens include *Usnea* species, while foliose types include *Xanthoria elegans* and *Umbilicaria*. Granite moss and two crustose lichens (*Buellia frigida* and *Rhizocarpon flavum*) may be found on rock surfaces.

The southern or continental zone is found in the southern and southeastern parts of the Peninsula, on adjacent islands, and on other limited coastal areas. Sea ice fringes most of the coast in winter, and some may not melt during summer. The warmest month here never averages above freezing and may be as low as 5° F (−15° C). Interior locations are colder, with January (summer) means of 10° F (−12° C). This zone is drier, with only 6 in (150 mm) of precipitation, almost all falling as snow. Although cold, the soil is too dry for patterned ground to form and also lacks an organic layer. The brief growing season, strong winds, wind-blown

· · · · · · · · · · · · · · · · · · · · · · · · · · · · · · · · · · · · ·

### Emperor Penguins

At more than 3 ft (1 m) tall and 88 lb (40 kg), Emperor Penguins are the largest in the world (see Plate VII). They eat fish, squid, and crustaceans captured during deep dives, reaching depths of 700 ft (200 m) or more lasting up to 20 minutes. Emperors breed on sea ice at the edge of the continent during winter and have no nests. Instead, they keep their eggs and young warm by perching them on their feet and insulating them with a special layer of skin called a brood pouch. After laying a single egg in May, the female returns to the sea, leaving the male to incubate the egg in total darkness, keeping it off the ice. When after two months (the incubation period) she returns, the male has lost up to one-third of his body weight and now must walk up to 60 mi (100 km) to the open sea to feed. After he returns, the pair alternate regurgitating meals for the chick and going back to the sea to feed. This takes place until January when the chicks are mature enough to walk to the ocean and begin independent lives. Ice conditions are important to successful breeding. If distance from the nesting site to the ocean is too far, the penguins may not have enough energy resources to feed the chick. If the sea ice breaks up too early in spring, the chicks may not be ready to survive in the open ocean.

· · · · · · · · · · · · · · · · · · · · · · · · · · · · · · · · · · · · ·

snow, long-lasting snow cover, and high salt content in the soil prohibit much plant growth. North-facing slopes have more vegetation because they receive more solar radiation, water from snowmelt or glaciers, and nutrients from bird or marine mammal droppings. Conditions are too severe for most organisms, but cyanobacteria, algae, lichens, and a few mosses survive. Liverworts are rare, and there are no flowering plants. Although most of the surface is bare, crustose lichens are found. Many lichens are endemic. Others are either cosmopolitan or also found in the Arctic. Under the best conditions, lichens grow only 0.4 in (1 cm) in 100 years. Cryptoendolithic lifeforms are microorganisms and crustose lichens living within pores of rocks.

No land mammals, birds, reptiles, or amphibians inhabit Antarctica, and no land mammals or birds breed south of South Georgia in the subantarctic. Terrestrial animals are limited to invertebrates such as rotifers, nematodes, water bears, mites, and springtails. As many as 50 seabird species, numbering in the millions, breed on the continent or islands in the summer, as do seals. Pelagic or far-ranging birds feed at sea and include albatrosses, fulmars, petrels, and shearwaters. Among them are five species of albatross, 10 different petrels, and both the Arctic and Antarctic Terns. Coastal species remain closer to shore and include skuas, cormorants, terns, and sheathbills. The skuas feed on land, stealing eggs from breeding sea-bird colonies. Because of the scarcity of snow-free ground, breeding takes place in large concentrations. Huge populations of marine organisms such as zooplankton, cephalopods, and fish support fur seals and southern elephant seals, which rest and breed on islands, especially Scotia Arc. In the summer, most coastal ice-free areas are intensely used by birds and seals, which transfer nutrients from sea to land in their waste.

Of the 17 species of penguin in the world, only four breed in Antarctica: Adelie, Emperor, Chinstrap, and Gentoo. Adelie and Emperor Penguins are most common. The others are found in the subantarctic islands, in southwest Africa, and even at the Equator in the Galapagos Islands. All penguins are unable to fly. Their shortened wings instead are used like paddles to propel them through water at speeds up to 25 mph (40 km/h). Their streamlined bodies have a thick layer of blubber as insulation and heavy bones to help keep them submerged in ocean water. Feathers are dense, forming a waterproof cover. Their diet is predominantly krill, taken from shallow water. They can leap several feet out of the water to get onto land or raised ice ledges. Although their waddling seems awkward, they can walk several miles over rocks, ice, or snow to ancestral nesting grounds called rookeries. If the surface is snow-covered, they save energy by sledding downhill. Some penguins are territorial, and their evenly spaced nests are a loose pile of small stones on snow-free rocky headlands. The normal clutch is two eggs, but only one chick may survive. Both males and females take turns incubating eggs and caring for chicks. When they are three to four weeks old, the chicks join nursery group huddles while both parents go to sea to feed. Parents feed their chicks by regurgitating into the beaks of their hungry offspring. While healthy adults have no land predators, several birds prey on eggs and young chicks.

Research stations and bases constructed in the last 100 years have destroyed much of the tundra, especially on the Antarctic Peninsula and adjacent islands. The delicate ecosystem is easily disrupted by pollution, research, tourism, and—potentially—mining.

## Further Readings

Antarctic Connection. n.d. http://www.antarcticconnection.com.

Arctic Portal. n.d. http://arcticportal.org/en/caff.

Dargaud, Guillaume. n.d. http://www.gdargaud.net/Antarctica/Penguins.html.

Environment Canada. n.d. http://www.mb.ec.gc.ca/nature/ecosystems/da00s04.en.html.

Ley, Willy. 1962. *The Poles, Life Nature Library.* New York: Time-Life.

National Oceanic and Atmospheric Administration. n.d. http://www.arctic.noaa.gov.

Polar Bears International. n.d. http://www.polarbearsinternational.org/about-us.

Smithsonian National Museum of Natural History. n.d. http://www.mnh.si.edu/arctic/html/wildlife.html.

# Appendix

## Biota of the Arctic and Antarctic Tundra Biome (arranged geographically)

### North American Arctic Tundra

#### Some Characteristic Plants of Low Arctic Tundra

*Treeline*

| | |
|---|---|
| White spruce | *Picea glauca* |
| Black spruce | *Picea mariana* |

*Tall shrubs (5–10 ft, 1.5–3 m)*

| | |
|---|---|
| Willow | *Salix alaxensis* |
| Mountain alder | *Alnus crispa* |
| Downy birch | *Betula pubescens* |

*Low shrubs (20 in, 50 cm)*

| | |
|---|---|
| Bog birch | *Betula nana* |
| Blue willow | *Salix glauca* |
| Tea-leaved willow | *Salix planifolia* |

*Heath family dwarf shrubs (4–8 in, 10–20 cm)*  Ericaceae

| | |
|---|---|
| Alpine bearberry | *Arctostaphylos alpina* |
| Black crowberry | *Empetrum nigrum* |
| Marsh Labrador tea | *Ledum palustre* |
| Bog bilberry or Alpine blueberry | *Vaccinium uliginosum* |
| Arctic bell heather | *Cassiope tetragona* |
| Mountain azalea | *Loiseleuria procumbens* |

*Graminoids*

| | |
|---|---|
| Nodding hairgrass | *Deschampsia flexuosa* |
| Matgrass | *Nardus stricta* |

| | |
|---|---|
| Red fescue grass | *Festuca rubra* |
| Arctic bluegrass | *Poa arctica* |
| Water sedge | *Carex aquatilis* |
| Sedge | *Carex lugens* |
| Few-flowered sedge | *Carex rariflora* |
| Bigelow's sedge | *Carex bigelowii* |
| Common cottongrass sedge | *Eriophorum angustifolium* |
| Tussock cottongrass sedge | *Eriophorum vaginatum* |
| Highland rush | *Juncus trifidus* |

### Forbs
*Cushions or mats*

| | |
|---|---|
| Arctic mountain avens | *Dryas integrifolia* |
| Mountain avens | *Dryas octopetala* |
| Moss campion | *Silene acaulis* |

*Leafy or rosettes*

| | |
|---|---|
| Viviparous bistort | *Polygonum viviparum* |
| Marsh cinquefoil | *Potentilla palustris* |
| Northern anemone | *Anemone parviflora* |

### Cryptogams

| | |
|---|---|
| Cetraria (Fruticose lichen) | *Cetraria* spp. |
| Reindeer moss (Fruticose lichen) | *Cladonia rangiferina* |
| Whiteworm lichen (Fruticose) | *Thamnolia vermicularis* |
| Aulacomnium (moss) | *Aulacomnium turgidum* |
| Haircap moss | *Polytrichum juniperinum* |
| Sphagnum moss | *Sphagnum* spp. |

## Some Characteristic Animals of Low Arctic Tundra

### Herbivores

| | |
|---|---|
| Caribou | *Rangifer tarandus* |
| Arctic hare | *Lepus arcticus* |
| Brown lemming | *Lemmus sibericus* |
| Collared lemming | *Dicrostonyx groenlandicus* |
| Arctic ground squirrel | *Spermophilus parryii* |

### Carnivores

| | |
|---|---|
| Brown or Grizzly bear | *Ursus arctos* |
| Timber or Arctic wolf | *Canus lupus* |
| Least weasel | *Mustela nivalis* |
| Short-tailed weasel or Ermine | *Mustela erminea* |
| Arctic fox | *Alopex lagopus* |

*Birds*

| | |
|---|---|
| Willow Ptarmigan | *Lagopus lagopus* |
| Snowy Owl | *Nystea scandiaca* |
| Pomerine Jaeger | *Stercorarius pomarinus* |
| Glaucous Gull | *Larus hyperboreus* |
| Short-eared Owl | *Asio flammeus* |
| Sandpiper | *Calidris* spp. |
| Lapland Larkspur | *Calcarius lapponicus* |
| Snow Bunting | *Plectrophenax nivalis* |
| Canada Goose | *Branta canadensis* |

## Some Characteristic Plants of High Arctic Tundra

*Low shrubs (20 in, 50 cm)*

| | |
|---|---|
| Arctic willow | *Salix arctica* |
| Arctic bell heather | *Cassiope tetragona* |
| Bog bilberry or Arctic blueberry | *Vaccinium uliginosum* |

*Graminoids*

| | |
|---|---|
| Mountain foxtail grass | *Alopecuris alpinus* |
| Fisher's dupontia grass | *Dupontia fisheri* |
| Rock sedge | *Carex rupestris* |
| Sedge | *Carex stans* |
| Short-leaf sedge | *Carex misandra* |
| White cottongrass sedge | *Eriophorum scheuchzeri* |
| Three-flowered rush | *Juncus albescens* |
| Northern woodrush | *Luzula confusa* |

*Forbs*

*Cushions or mats*

| | |
|---|---|
| Arctic mountain avens | *Dryas integrifolia* |
| Red sandwort | *Minuartia rubella* |
| Tufted saxifrage | *Saxifraga caespitosa* |
| Nodding saxifrage | *Saxifraga cernua* |
| Purple saxifrage | *Saxifraga oppositifolia* |
| Starwort | *Stellaria* spp. |

*Leafy or rosettes*

| | |
|---|---|
| Flat-top draba | *Draba corymbosa* |
| Lousewort | *Pedicularis* spp. |
| Dwarf willow | *Salix herbaceae* |
| Rooted poppy | *Papaver radicatum* |
| Mouse-ear | *Cerastium alpinum* |

*Cryptogams*

| | |
|---|---|
| Cetraria (Fruticose lichen) | *Cetraria* spp. |
| Arctic finger lichen (Fruticose) | *Dactylina arctica* |

| | |
|---|---|
| Snow lichen (Fruticose) | *Stereocaulon alpinum* |
| White worm lichen (Fruticose) | *Thamnolia vermicularis* |
| Dermatocarpon (Crustose lichen) | *Dermatocarpon hepaticum* |
| Lecanora (Crustose lichen) | *Lecanora epibyron* |
| Ditrichum (moss) | *Ditrichum flexicaule* |
| Hylocomium (moss) | *Hylocomium splendens* |
| Haircap moss | *Polytrichum juniperinum* |
| Tomenthpynum (moss) | *Tomenthpynum nitens* |

## Some Characteristic Animals of High Arctic Tundra

### Herbivores
| | |
|---|---|
| Peary's caribou | *Rangifer tarandus pearyi* |
| Muskox | *Ovibus moschatus* |
| Collared lemming | *Dicrostonyx groenlandicus* |

### Carnivores
| | |
|---|---|
| Polar bear | *Ursus maritimus* |
| Timber or Arctic wolf | *Canis lupus* |
| Short-tailed weasel or Ermine | *Mustela erminea* |
| Arctic fox | *Alopex lagopus* |

### Birds
| | |
|---|---|
| Rock Ptarmigan | *Lagopus mutus* |
| Snowy Owl | *Nystea scandiaca* |
| Pomerine Jaeger | *Stercorarius pomarinus* |
| Glaucous Gull | *Larus hyperboreus* |
| Snow Goose | *Chen caerulescens* |

## Some Characteristic Plants of Greenland Arctic Tundra

### Low shrubs (20 in, 50 cm)
| | |
|---|---|
| Downy birch | *Betula pubescens* |
| Greenland ash | *Sorbus groenlandica* |
| Arctic willow | *Salix arctica* |
| Blue willow | *Salix glauca* |
| Arctic bell heather | *Cassiope tetragona* |
| Bog bilberry | *Vaccinium uliginosum* |
| Mountain azalea | *Loiseleuria procumbens* |
| Alpine bearberry | *Arctostaphylos alpina* |

### Graminoids
| | |
|---|---|
| Arctic bluegrass | *Poa arctica* |
| Bigelow's sedge | *Carex bigelowii* |
| Common cottongrass sedge | *Eriophorum angustifolium* |

*Forbs*
*Cushions or mats*
Mountain avens                                    *Dryas octopetala*
Purple saxifrage                                  *Saxifraga oppositifolia*
Moss campion                                      *Silene acaulis*

*Leafy or rosettes*
Viviparous bistort                                *Polygonum viviparum*

## Some Characteristic Animals of Greenland Arctic Tundra

**Herbivores**
Caribou                                           *Rangifer tarandus*
Muskox                                            *Ovibus moschatus*
Arctic hare                                       *Lepus arcticus*
Collared lemming                                  *Dicrostonyx torquatus*

**Carnivores**
Timber or Arctic wolf                             *Canis lupus*
Arctic fox                                        *Alopex lagopus*
Short-tailed weasel or Ermine                     *Mustela erminea*
Polar bear                                        *Ursus maritimus*

**Birds**
Wheatear                                          *Oenanthe oenanthe*
Snow Bunting                                      *Plectrophenax nivalis*
Greenland Redpoll                                 *Carduelis flammea rostrata*
Lapland Bunting                                   *Calcarius lapponicus*
Rock Ptarmigan                                    *Lagopus mutus*
Snowy Owl                                         *Nystea scandiaca*
Arctic Redpoll                                    *Carduelis flammea hornemanni*
Pink-footed Goose                                 *Anser brachyrhynchus*
Barnacle Goose                                    *Branta leucopsis*

## Eurasian Arctic Tundra

## Some Characteristic Plants of Fennoscandia Arctic/Alpine Tundra

**Treeline**
Downy birch                                       *Betula pubescens*
Norway spruce                                     *Picea abies*

**Low shrubs (8 in to 10 ft, 20 cm to 3 m)**
Bog birch                                         *Betula nana*
Common juniper                                    *Juniperus communis*
European blueberry                                *Vaccinium myrtillus*
True or Scotch heather                            *Calluna vulgaris*

| Bog rosemary | *Andromeda polifolia* |
| Small cranberry | *Vaccinium oxycoccus* |
| Moss plant | *Cassiope hypnoides* |

### Graminoids

| Nodding hairgrass | *Deschampsia flexuosa* |
| Matgrass | *Nardus stricta* |
| Sheep fescue grass | *Festuca ovina* |
| Spiked trisetum grass | *Trisetum spicatum* |
| Tussock cottongrass sedge | *Eriophorum vaginatum* |
| Bigelow's sedge | *Carex bigelowii* |
| Highland rush | *Juncus trifidus* |

### Forbs

*Cushions and mats*

| Nodding saxifrage | *Saxifraga cernua* |
| Tufted saxifrage | *Saxifraga caespitosa* |
| Purple saxifrage | *Saxifraga oppositifolia* |
| Moss campion | *Silene acaulis* |

*Leafy forbs*

| Monkshood | *Aconitum septentrionale* |
| Globeflower | *Trollius europaeus* |
| Dwarf willow | *Salix herbacea* |
| Buttercup | *Ranunculus platanifolius* |
| Alpine sorrel | *Oxyria digyna* |

### Cryptogams

| Mountain lichen (Fruticose) | *Solorina corceae* |
| Iceland moss (Fruticose lichen) | *Cetraria islandica* |
| Tree reindeer lichen (Fruticose) | *Cladonia mitis* |
| Sphagnum moss | *Sphagnum* spp. |

## Some Characteristic Animals of Fennoscandia Arctic/Alpine Tundra

### Herbivores

| Reindeer | *Rangifer tarandus* |
| Muskox | *Ovibus moschatus* |
| Norway lemming | *Lemmus lemmus* |
| Tundra or Root vole | *Microtus oeconomus* |
| Mountain hare | *Lepus timidus* |

### Carnivores

| Common shrew | *Sorex araneus* |
| Lynx | *Felis lynx* |

(*Continued*)

| Short-tailed weasel or Ermine | *Mustela erminea* |
| Least weasel | *Mustela nivalis* |
| Red fox | *Vulpes vulpes* |

### *Birds*

| Whooper Swan | *Cygnus cygnus* |
| Bean Goose | *Anser fabalis* |
| Mallard | *Anas platyrhynchos* |
| Golden Plover | *Pluvialis apricaria* |
| Red-necked Phalarope | *Phalaropus lobatus* |
| Snow Bunting | *Plectrophenax nivalis* |
| Meadow Pipit | *Anthus pratensis* |
| Wheatear | *Oenanthe oenanthe* |
| Willow Ptarmigan | *Lagopus lagopus* |
| Rock Ptarmigan | *Lagopus mutus* |
| Long-tailed Skua | *Stercorarius longicaudus* |
| Snowy Owl | *Nyctea scandiaca* |
| Short-eared Owl | *Asio flammeus* |
| Rough-legged Buzzard | *Buteo lagopus* |
| Golden Eagle | *Aquila chrysaetos* |
| Gyrfalcon | *Falco rusticolus* |
| Raven | *Corvus corax* |

### *Reptiles and amphibians*

| European common frog | *Rana temporaria* |
| Common viviparous lizard | *Lacerta vivipara* |
| Grass snake | *Natrix natrix* |

## Some Characteristic Plants of Russian Arctic Tundra

### *Treeline*

| Downy birch | *Betula tortuosa* |
| Chosenia | *Chosenia arbutifolia* |
| Siberian larch | *Larix siberica* |
| Siberian spruce | *Picea obovata* |
| Mongolian poplar | *Populus suaveolens* |

### *Tall shrubs (2 ft, 0.6 m)*

| Resin birch | *Betula glandulosa* |
| Bog birch | *Betula nana* |
| Felt-leaf willow | *Salix alaxensis* |
| Blue willow | *Salix glauca* |
| Alder | *Alnaster fruticosus* |
| Mountain alder | *Alnus crispa* |

### *Dwarf shrubs (6 in, 15 cm)*

| Arctic bell heather | *Cassiope tetragona* |

| | |
|---|---|
| Mountain crowberry | *Empetrum hermaphroditum* |
| Bog bilberry | *Vaccinium uliginosum* ssp. |
| | *microphyllum* |
| Polar willow | *Salix polaris* |

### Graminoids
| | |
|---|---|
| Fisher's dupontia grass | *Dupontia fisheri* |
| Siberian spike-seed grass | *Trisetum sibiricum* |
| Tufted hairgrass | *Deschampsia caespitosa* |
| Arctic bluegrass | *Poa arctica* |
| Carex sedge | *Carex lugens* |
| Common cottongrass sedge | *Eriophorum angustifolium* |
| Tufted cottongrass sedge | *Eriophorum vaginatum* |
| Northern woodrush | *Luzula confusa* |

### Forbs
*Cushions or mats*
| | |
|---|---|
| Mountain avens | *Dryas octopetala* |
| Tufted saxifrage | *Saxifraga caespitosa* |
| Hawkweed-leaved saxifrage | *Saxifraga hieracifolia* |
| Arctic saxifrage | *Saxafraga hirculus* |
| Polar poppy | *Papaver polare* |

*Leafy or rosettes*
| | |
|---|---|
| Alpine sorrel | *Oxyria digyna* |
| Snow buttercup | *Ranunculus nivalis* |
| Sulphur-colored buttercup | *Ranunculus sulphureus* |
| Long-stalked starwort | *Stellaria longipes* |

### Cryptogams
| | |
|---|---|
| Iceland moss (Fruticose lichen) | *Cetraria islandica* |
| Gray reindeer lichen (Fruticose) | *Cladina rangiferina* |
| Arctic finger lichen (Fruticose) | *Dactylina arctica* |
| Snow lichen (Fruticose) | *Stereocaulon alpinum* |
| Whiteworm lichen (Fruticose) | *Thamnolia vermicularis* |
| Northern reindeer lichen (Fruticose) | *Cladina stellaris* |
| Green dog lichen (Foliose) | *Peltigera aphthosa* |
| Parmelia (Foliose lichen) | *Parmelia omphalodes* |
| Biatora (Crustose lichen) | *Biatora* |
| Lecidea (Crustose lichen) | *Lecidea* |
| Ochrolechia (Crustose lichen) | *Ochrolechia* |
| Aulacomnium (moss) | *Aulacomnium turgidum* |
| Hylocomium (moss) | *Hylocomium alaskanum* |
| Tomenthypnum (moss) | *Tomenthypnum nitens* |
| Giant calliergon moss | *Calliergon giganteum* |

*(Continued)*

| Drepanocladus (moss) | *Drepanocladus uncinatus* |
| Sphagnum moss | *Sphagnum* spp. |
| Tritomaria (liverwort) | *Tritomaria quinquedentata* |
| Ptilidium (liverwort) | *Ptilidium ciliare* |

## Some Characteristic Animals of Russia Arctic Tundra

### Herbivores

| Reindeer | *Rangifer tarandus* |
| Muskox | *Ovibus moschatus* |
| Amur lemming | *Lemmus amurensis* |
| Brown lemming | *Lemmus sibericus* |
| Tundra or Root vole | *Microtis oeconomus* |

### Carnivores

| Polar bear | *Ursus maritimus* |
| Brown or Kamchatka bear | *Ursus arctos* |
| Arctic fox | *Alopex lagopus* |
| Short-tailed weasel or Ermine | *Mustela erminea* |

### Birds

| Willow Ptarmigan | *Lagopus lagopus* |
| Rock Ptarmigan | *Lagopus mutus* |
| Stone Plover | *Limosa lapponica* |
| Rough-legged Buzzard | *Buteo lagopus* |
| Arctic Skua | *Stercorarius parasiticus* |
| Snowy Owl | *Nyctea scandiaca* |
| Snow Bunting | *Plectrophenax nivalis* |
| White-fronted Goose | *Anser albifrons* |
| Bean Goose | *Anser fabalis* |
| Brent Goose | *Branta bernicola* |
| Snowy Owl | *Nyctea scandiaca* |
| Curlew Sandpiper | *Calidris ferruginea* |

## Antarctic Tundra

## Some Characteristic Plants of Antarctic Tundra

### Graminoids
| Antarctic hairgrass | *Deschampsia antarctica* |

### Forbs (cushion plant)
| Antarctic pearlwort | *Colobanthus quitensis* |

### Cryptogams
| Usnea (Fruticose lichen) | *Usnea aurantiacoatra* |
| Usnea (Fruticose lichen) | *Usnea antarctica* |

| | |
|---|---|
| Xanthoria (Foliose lichen) | *Xanthoria elegans* |
| Umbilicaria (Foliose lichen) | *Umbilicaria decussata* |
| Umbilicaria (Foliose lichen) | *Umbilicaria aprina* |
| Buellia (Crustose lichen) | *Buellia frigida* |
| Rhizocarpon (Crustose lichen) | *Rhizocarpon flavum* |
| Haircap moss | *Polytrichum alpestre* |
| Chorisodontium (moss) | *Chorisodontium aciphyllum* |
| Drepanocladus (moss) | *Drepanocladus* |
| Brachythecium (moss) | *Brachythecium* |
| Acrocladium (moss) | *Acrocladium* |
| Granite moss | *Andreaea* |

## Some Characteristic Animals of Antarctic Tundra

### Sea mammals

| | |
|---|---|
| Fur seal | *Arctocephalus gazella* |
| Southern elephant seal | *Mirunga leonina* |

### Birds

| | |
|---|---|
| Adelie Penguin | *Pygoscelis adelias* |
| Emperor Penguin | *Aptenodytes forsteri* |
| Gentoo Penguin | *Pygosceles papua* |
| Chinstrap Penguin | *Pygosceles antarctica* |
| Antarctic Petrel | *Thalassoica antarctica* |
| Wilson's Storm Petrel | *Oceanites oceanicus* |
| Antarctic Skua | *Catharacta antarctica* |
| Wandering Albatross | *Diomedea exulans* |
| Royal Albatross | *Diomedea epomophora* |
| Snow Petrel | *Pagodroma nivea* |
| Arctic Tern | *Sterna paradisae* |
| Antarctic Tern | *Sterna vittata* |
| Southern Fulmar | *Fulmarus glacialoides* |

# 3

..............................................................................

# Mid-Latitude Alpine Tundra

The term mid-latitude refers to the areas between the Tropic of Cancer and the Arctic Circle and the Tropic of Capricorn and the Antarctic Circle—the regions between $23^1/_2°$ and $66^1/_2°$ in both hemispheres. Mid-latitudes in general are characterized by seasonality of temperature due to a marked change in sun angle and length of day from summer to winter. Mid-latitude alpine tundra is the life zone on mountains above treeline. Alpine tundra of most Eurasian and North American mountains is similar to that of the Arctic but with a richer flora. Limited in extent in the Southern Hemisphere, mid-latitude alpine life zones have growthforms and environments similar to those in the north, although the plant species present are quite different. Small areas of alpine environment on mountain summits are difficult to illustrate on a world map.

Although it is often stated that altitude mimics latitude, many important differences exist between arctic and alpine environments. Both regions have low mean annual temperature and short growing seasons. Both experience a seasonal change in sun angle—higher in the summer and lower in the winter—which contributes to seasonality of temperature. The light regime differs, however. Alpine regions never experience 24 hours of day or night, and duration of daylight or darkness varies according to latitude. Although permafrost occurs on high mountains, arctic landforms do not occur because few large, flat areas on which they could develop exist. Disturbances like avalanches and rock slides are common in steep mid-latitude mountains but are rare to nonexistent on level expanses in the Arctic.

## Physical Environment

### Mid-Latitude Alpine Climates

All alpine areas have climate characteristics in common, many of which are inter-related. Most important are atmospheric pressure, temperature, and moisture content of the air. The density of air decreases as altitude increases. At 8,550 ft (2,600 m), air pressure is 26% lower than at sea level, and at 19,000 ft (5,800 m), which is close to the upper limit of vascular plant life in the Himalayas and the Alps, it is 50% lower. Less dense air means less carbon dioxide for plants to use in photosynthesis. Temperature decreases with elevation at an average rate of 3.5° F for every 1,000 ft of elevation (6.5° C per 1,000 m), known as the average lapse rate. Actual lapse rates, however, vary depending on inland or coastal location, latitude, and weather. In spite of sometimes high relative humidity readings, alpine air is usually very dry. Little water vapor can be held in the thin air, and low temperatures inhibit evaporation. Temperatures at ground level, however, are higher, which increases both evaporation and relative humidity.

Few long-term climate records exist for mid-latitude areas above treeline. In general, soil and air temperatures are low, and the growing season is 4–10 weeks long. Niwot Ridge west of Boulder, Colorado, is typical of alpine climates in continental areas (see Figure 3.1). At standard weather station height of 5 ft (1.5 m), the mean annual air temperature at 12,300 ft (3,750 m) is 25° F (−3.7° C) and mean July temperature is 46° F (8° C). Temperatures in the Sierra Nevada are slightly warmer because of a dry summer, while alpine zones in maritime regions such as Mount Rainier in the Cascade Range have both more precipitation and moderate temperatures. Temperatures closer to ground level and extremes during the short growing season are more important for plant life than average temperatures. Plants 8 in (20 cm) high experience July temperatures ranging from 70° F (21° C) during the day down to 39° F (4° C) at night. Soil temperatures to a depth of 8 in (20 cm) are similar to the air above. Even without considering extremes, ground-level climates have a temperature range of 30° F (16° C) from day to night. Although low temperatures are common during the mid-latitude growing season, freezing temperatures are rare.

*Macro-, meso-, and microclimates.* It is hard to generalize a mountain climate, and conventional climate statistics and descriptions are misleading. Alpine climates can be described on three scales. Macroclimate refers to the general climate of the geographic location, such as maritime for the North American Cascades and European Alps, continental for the Rocky Mountains, or semidesert for the Great Basin of North America and Tibet. Mesoclimate takes into account variation with elevation, slope steepness, aspect, wind, and water, all of which are interrelated. For example, a north-facing slope in the Northern Hemisphere will be shadier and more humid, allowing forest, and consequently treeline, to grow at a higher elevation than on the hotter and drier south-facing slope.

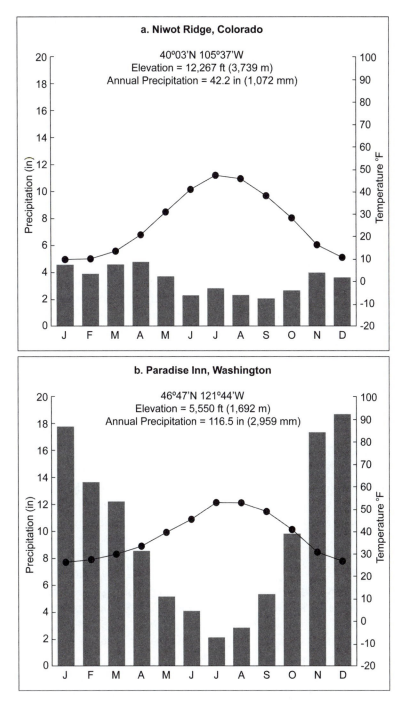

**Figure 3.1** (a) Niwot Ridge in the Southern Rocky Mountains of Colorado is typical of mid-continental alpine tundra. (b) The higher precipitation and mild winter temperatures of Paradise Inn on Mount Rainier in Washington illustrate a maritime influence. *(Illustration by Jeff Dixon.)*

Rugged topography creates a mosaic of microclimates (climate in a small area), which may change every few feet. Because meso- and microenvironments may be different from macroclimate, elevation is not always the best criterion for plant growth limits. Microclimate on a slope might override climate conditions of elevation, so that a plant may be found higher than what is considered normal. Both rocks and vegetation clumps affect wind and snow patterns. Snow often accumulates in the lee of the obstruction, altering the microhabitat there. An east-facing niche in the lee of a boulder on a windy west slope, for example, may accumulate enough snow to protect a plant from temperature extremes. The highest vascular plant in the world, snow lotus (*Saussurea gnaphalodes*), can grow on scree at 21,000 ft (6,400 m) in the Himalayas because the microclimate there is similar to conditions at 13,000 ft (4,000 m) where it normally grows. The three most important microclimatic factors are solar radiation, slope steepness, and slope aspect or exposure (direction the slope faces). Others include wind speed, air temperature above vegetation, and soil characteristics of structure, temperature, and moisture. At any scale, a change in one of these factors may affect many others. For example, different amounts of incoming solar radiation will cause differences in soil temperature, air temperature, snow cover duration, and soil moisture, all of which affect vegetation. Microclimate diversity permits high biological diversity.

**Solar Collectors**

Energy from the sun is most efficiently absorbed when the surface is at a 90-degree angle to the sun's rays. Solar collectors never lie flat on the ground, but instead are inclined at an angle to intercept the sun's rays as directly as possible. Because sun angle varies both with latitude and season of the year, inclination of the collectors must be aligned according to latitude and be able to move seasonally as the sun moves higher or lower in the sky. For even more efficiency, they might also adjust throughout the day to follow the sun's path through the sky from sunrise to sunset. Mountain slopes in the alpine zone cannot adjust to variation in the incoming solar rays, but plants have adapted to the different habitats afforded by different slopes and aspects.

***Radiation and temperature.*** Because of the thin atmosphere at high elevations, alpine areas are often subjected to intense solar radiation. Sunny and shady microclimates experience very different temperatures. The shady area measures actual air temperature in the thin atmosphere, while the sunny area is directly warmed by the sun's rays. Diurnal temperature changes can be extreme because of the thin atmosphere. Intense solar radiation reaches the surface and is absorbed by the ground, which becomes warm. At night, clear skies allow much infrared radiation to escape back through the atmosphere and return to space, making nights cold. Increased cloudiness limits the passage of both solar and infrared radiation, narrowing the temperature difference between both day to night and microclimates. A snow cover will reflect much of the solar radiation during the day and also insulate the ground, preventing terrestrial radiation from escaping at night. Both processes reduce temperature and diurnal temperature ranges.

*Slope angle and aspect.* The steepness of a slope affects the angle at which sunlight strikes the surface. Even at a high latitude, a very steep slope might receive close to a 90-degree sun angle, increasing the amount of solar energy it absorbs. The direction of exposure adds another dimension to temperature variation. Slope aspect affects the time of day or year sunlight can reach the surface. In mid-latitudes of the Northern Hemisphere, the south-facing slope is sunny during all daylight hours in both summer and winter. The north-facing slope will only get sunlight on summer mornings and evenings, and in winter will receive none. The opposite is true for the Southern Hemisphere: the north-facing slope is sunny while the south-facing slope is in shadow. An extreme example of temperature differences according to slope aspect was measured in the Austrian Alps. Surface temperature on the southwest slope in summer was 176° F (80° C), while the northeast slope was only 73° F (23° C). Sunny slopes experience more temperature change from day to night, and snow melts sooner on sunny slopes.

### Surface Temperatures

Because the ground more efficiently absorbs and loses heat than the air does, difference in surface temperatures from day to night can be extreme. That is a major reason why sand on the beach is hot during the day. The ground absorbs solar radiation all day, concentrating the heat at the surface. It is then readily available to be radiated back to space at night, causing the surface to be cold. Horizontal surfaces lose more radiation at night than vertical surfaces lose. Frost may form on the sloping windshield of your vehicle but not on the vertical side windows because more radiation is lost from the more-horizontal surface, making it colder.

*Wind.* In mid-latitudes, strong winds are usually associated with a position under the jet stream or in the paths of cyclonic storms. Some mountains are windy while others are not. Windiness is more related to topography than to elevation, being strongest on exposed ridges. Winds are weaker at ground level where low-stature plants grow. Wind redistributes snow, affecting the moisture gain or loss for an area. Strong winds, especially in dry air, encourage sublimation, preventing water gain from snowmelt, and also increase evapotranspiration from plants, both lowering their internal temperatures and causing them to dry out. Wind affects heat exchange by mixing layers of air, cooling the surface during day and warming it at night.

In rugged topography, upslope and downslope winds alternate between day and night. Intense warming during the day creates a low pressure, causing air to rise. The resulting upslope winds frequently give rise to thunderstorms in the afternoon. After sunset, the alpine air rapidly cools, creating a high pressure that causes air to sink into the valleys below. Downslope winds at night, called cold-air drainage, can cause temperature inversions by displacing warm air, making valleys cooler than the alpine environments at higher elevations. Such inversions may be a nightly occurrence, literally turning the cooler-with-elevation norm upside down, and can contribute to the development of microclimates.

*Precipitation.* Precipitation in mid-latitude mountains generally increases with elevation all the way up to the summits. The potential for evapotranspiration decreases with elevation, primarily because of cooler temperatures and shorter growing seasons. Low temperatures and low evaporation indicate a wet climate with plenty of moisture. Therefore, it would appear that most alpine areas should have an abundance of moisture, at least on an annual basis. However, that relationship is misleading because all precipitation may not become available for plant use. As much as 80% of that which falls in alpine areas ends up as runoff from snowmelt or percolates through soil and rocks, eventually to reemerge in rivers and streams. Therefore, alpine zones are often considered arid, a concept reinforced by widely dispersed vegetation. Furthermore, the higher temperatures experienced by small alpine plants close to the ground result in higher evapotranspiration during the growing season, often resulting in a water shortage by summer's end. Low annual precipitation totals recorded in some mid-latitude alpine habitats may be artifacts of measurement errors caused by few weather stations, strong winds, inaccurate determination of the water equivalence of snowfall, or evaporation.

Snow protects plants from both wind and cold, and different plant communities grow on windy snowfree ridges than those found in snow-filled gullies. Lack of snow exposes plants to daily extremes of full solar radiation and nighttime infrared loss. Measurements of microclimate temperatures in the Alps during winter show that plants can be exposed to diurnal temperatures as high as 86° F (30° C) and as low as 14° F (−10° C). Wind and accompanying frigid temperatures often sculpt or prune any parts of plants that project above snow level as evidenced by krummholz at treeline. However, plants also have an effect on wind, decreasing its speed and causing snow drifts to accumulate on their lee sides.

## Soils

Because of low soil temperatures and uneven distribution of soil moisture, soil development is slow. Microorganism activity is especially inhibited. Most alpine soils are regularly disturbed by freezing, erosion, and downslope slippage and contain a mix of large rocks and fine particles. Large rocks derive from onsite weathering, glacial deposits, and landslides. Sources of fine material include weathering of parent rock in place and from water, avalanche, and wind deposition. Dust, blown up from adjacent lowlands or blown in from nearby glacial sediments, can add nutrients not present in local rocks. Soils typically vary over short distances—sand, peat bogs, shallow carbonate soils, and deep profiles in alpine grasslands—and the variation contributes to microhabitat development. Wind-blown material may be a major component of grassland soils where limited water flow fails to bring in sediment. Cushion plants, abundant where it is windy, trap sediment and nutrients, enriching both plant and soil. Fine material accumulates in spaces between large rocks, where plants subsequently germinate and root. If soil, however, is buried too deep below the rocky surface, as commonly happens on scree slopes, it fails to become warm enough to support most plants. Except in meadows and bogs, organic matter is extremely limited in alpine soils because plant cover is sparse.

Frost action, either diurnal or seasonal, disrupts soil formation, making it hard for plants to colonize. Needle ice occurs almost nightly during the growing season, uprooting small plants and churning up seed beds. At a much larger scale on a seasonal basis, frost-heaving can create patterned ground and solifluction. Most large-scale patterned ground in the alpine zone, however, was formed during the Pleistocene and is no longer active.

At the beginning of the growing season, most soils are saturated from snowmelt, and areas where snow was deep or runoff accumulates may be saturated all summer. Later in summer, after the snow has melted, soils tend to dry out. However, plants are rarely subjected to drought stress. While the top inch (2.5 cm) of soil may dry out even in the wettest mountains, deeper soils retain water that can be tapped by deep-rooted plants. The Pamirs in the interior of Asia are one of the driest alpine regions. Yet, even with less than 12 in (300 mm) of annual precipitation, little snow cover, surface desiccation, and low relative humidity, vegetation in the alpine semidesert at 14,000 ft (4,250 m) is not drought-stressed. Measurements of leaf moisture loss and transpiration rates indicate that plants are getting sufficient soil moisture.

Three major categories of soils occur in alpine areas. Poorly developed rocky or stony soils (or entisols) are called lithosols when formed on bare rock of slopes and ridges and regosols when formed on unconsolidated rock like talus or scree. Parent material is the only difference, and both types are shallow, dry, and azonal, meaning no defined horizons or layers are apparent. Plant communities on entisols are usually dominated by lichens, mosses, and cushion plants. Alpine turf and meadow soils (or inceptisols) are up to 30 in (75 cm) deep and exhibit distinct horizons. The dark A horizon contains finer soil particles and has considerable organic content, while the B horizon is lighter in color because it contains less humus. Both turf and meadow soils support herbs and grasses with their mesh of intertwined roots. Bog soils (or histosols) are found in depressions or where seepage saturates ground to depths of 3.3 ft (1 m). Mossy bogs often overlie and insulate permafrost,

## Succulents in Alpine Tundra

Succulent plants are found in alpine regions of all climate zones. Both in the crassula family, mountain houseleek grows at 10,500 ft (3,200 m) in the Alps, and yellow stonecrop occurs at 12,250 ft (3,725 m) in the Rocky Mountains. Also a member of the crassula family, a type of hen and chicks, *Echeveria columbiana* is found at 13,500 ft (4,100 m) in Venezuela. Cacti also grow high in the Andes. Several species of *Tephrocactus*, such as paper-spine cactus, which has long, flat, flexible spines, occur above 14,000 ft (4,250 m). Several alpine cacti, such as *Opuntia lagopus* at 14,750 ft (4,500 m) in Peru, and many others called old man cactus throughout the high Andes, have protective coverings of long white spines or hairs. $C_3$ plants dominate in alpine environments, and plants dependent on $C_4$ or Crassulacean Acid Metabolism (CAM) methods are rare. Treeline is generally the upper limit of most $C_4$ plants, and succulent CAM plants are restricted to hot or dry microhabitats. Most alpine succulents deal with surface drought by storing water.

which contributes to the poor drainage conditions. In the saturated soils, a lack of oxygen inhibits decay of plant matter and makes bog soils acidic.

## Plant Adaptations

Alpine growthforms are predominantly perennial herbs, graminoids, rosettes, cushions, and low shrubs. Graminoids (grasses and sedges) are more typical of moist habitats, while rosettes and cushions occur on windy ridges or areas that lack protective snow cover. Cushions are prominent where low soil temperatures inhibit elongation of stems. Small, compact leaves help maintain higher temperatures in the plants. In an experiment growing the same plants at different elevations, individuals in the alpine zone were shorter and produced fewer branched flower stems as well as fewer and smaller leaves than individuals of the same species grown at lower elevations. Prostrate woody shrubs, either evergreen or deciduous are common, as are lichens and bryophytes. The alpine flora of any given mountain region commonly has 200–300 kinds of higher plants. Most have temperate origins and are mainly members of the rose, pink, buckwheat, mustard, and saxifrage families. Most perennials have more biomass in roots than in leaves and shoots, because roots are essential to anchor the plant, absorb water and nutrients, and store the carbohydrates necessary for rapid growth in spring. Annuals make up less than 6% of the total alpine flora of most mid-latitude mountains; they need moist soil all summer and bare sites with little competition, an unlikely combination in the alpine environment.

Most alpine plants can begin growing at temperatures just above freezing, while lowland plants usually require temperatures of 40°–55° F (4°–13° C). Early summer growth draws on energy supplies stored in the root or tuber, which are then replaced during the summer. Many plants have a red color due to anthocyanin pigments in stems and leaves. Anthocyanins convert light to heat, which is especially important in the spring and an important factor in making plants cold hardy. The red color shows particularly at the beginning and end of the growing season; it is masked by green chlorophyll in summer. Hairs on leaves or buds offer protection from strong solar radiation and also conserve heat. Many plants have a natural "antifreeze" that keeps the tissue from freezing when air temperature is below freezing.

Under deep snowpacks that limit the growing season, some plants preform flower and leaf buds at the end of the previous summer season. Plants may have microhabitat preferences around a snowbed, related to the average date of melting. In the Rocky Mountains, for example, alpine avens, creeping sibbaldia, and Rocky Mountain sage grow at the outer edges of a snowbed, while snow buttercup, which tolerates a shorter season, grows toward the center. Zones where snow melts faster not only have longer growing seasons but also more nutrient-rich soils. Humus content is higher because plants grow for more summer days. Snowbeds are also good sources of soluble nutrients such as potassium and calcium, which accumulate over the winter and are released as the snow melts. However, the surface beneath extremely late-lying snowbeds may be bare or support only mosses and lichens.

Because most alpine plants are not constrained by water shortages, desiccation is rare and only found in specific habitats. Plants have various means by which to obtain moisture. Cushions, abundant in both humid microclimates and exposed ridges, are commonly deep-rooted, and also trap water in their compact canopies. Many lichens absorb their own weight in water, almost directly into the thallus cells, a much quicker process than via roots and stems. Mosses also absorb water like a sponge.

At the upper limit of vascular plant growth, small rosettes and cushions are common growthforms. Wind-swept ridges support plants like saxifrages similar to those found in the Arctic, but because more rocky crevices are available to trap snow and provide more water, soft-tissued plants like buttercup also thrive. In wet protected pockets called "snow crannies," buttercups can grow at higher elevations than even cushion plants. Glacier buttercup at 14,000 ft (4,275 m) is the highest vascular plant in the Alps, and Graham's buttercup in New Zealand grows in the permanent snow zone at 9,500 ft (2,900 m). In high-elevation snow crannies, snowmelt provides moisture in otherwise dry sites while the surrounding rocks absorb and radiate heat.

## Reproduction

Alpine plants have three ways of maintaining their position within the community, two by reproduction and one by being long-lived. Most plants reproduce sexually even though conditions may not be favorable for completion of the cycle every year. Vegetative reproduction is also important for most plants. Plants are slow growing and may be 10–15 years old before flowering.

The three patterns of flowering times—early summer, mid-summer, and late summer—overlap in mid-summer because of the shortness of the season. Early-flowering plants do so at or shortly after the snow melts, mid-season plants flower at the peak of vegetative growth, and flowers of late-season plants appear late in the growing period after all leaves have been produced. All early- and mid-season plants exhibit flower preformation. Inflorescences begin development in the previous season; how much is preformed depends on how soon in the summer season the plant flowers. Glacier buttercup flowers early and may have two to three generations of preformed flowers blooming in a single season. In most mid-latitude mountains, peak flowering is mid to late July. Late-season plants usually do not have preformed flowers. Viviparous bistort, whose preformed flowers finish blooming late in the summer, is an exception. It requires three seasons to form the bud and does not flower until the fourth.

Temperature and photoperiod (hours of daylight) are two factors that exert the most control over the time of flowering. Grasses such as alpine bluegrass must undergo cold winter temperatures. Many plants need a long day in addition to one

···············································

### Seeds

Alpine plants produce a lot of seeds. Average output from one buttercup plant is 500, and one alpine willowherb individual will produce 60,000 small plumed seeds. In Alaska, 1,155 viable seeds were recovered in one square meter of soil in alpine tundra. Such high seed production usually is a hedge against high potential mortality.

···············································

or two cold winters. Very early bloomers, though, such as primrose, buttercups, purple saxifrage, and most carex sedges and woodrushes, are more dependent on time of snowmelt than photoperiod. Late-season flowers such as alpine bluegrass and alpine timothy respond to decreasing photoperiod in late summer. Alpine flower displays are known for their mosaics of colors and patterns.

Two different sexual reproduction strategies are common and vary with respect to the number of seeds produced and the type of pollination undergone. Under favorable conditions, seeds are abundant. Early-flowering species produce fewer seeds but have a greater chance that their seeds will mature and ripen. Late-flowering plants produce more seeds to increase the chance that some might survive to maturity. Early-season plants are more likely to be cross-pollinated, while more late-season plants depend on self-pollination. The number of pollinators does not decrease with elevation, although the spectrum changes. Butterflies and beetles are replaced by bumblebees and flies. Bumblebees, however, are immobilized at temperatures below 50° F (10° C). Above 13,200 ft (4,000 m) in the Himalayas, bee-pollinated flowers cannot grow, and plants depend on other insects. In the Alps, 29 different insect families were observed visiting edelweiss. Hummingbirds are important pollinators in some warm-summer mid-latitude mountains like the Sierra Nevada and southern Rockies, but none occur in polar mountains. Wind can be important in dispersing pollen from one plant to another.

Germination and seedling establishment are difficult on surface soils prone to desiccation and needle-ice disturbance. To survive, the seed must become firmly anchored in deeper soil. Some seeds survive in crevices or in shelter of existing plants, especially cushion canopies. Germination can be extremely rapid, within one week after snowmelt. Few plants have seeds that require a dormant period. Delayed germination until the following summer is environmentally imposed by low soil moisture, ensuring that the plant will not germinate too late in the short growing season. Seeds of plants growing in wet meadows, such as glacier lily, snowball saxifrage, tufted hairgrass, and American bistort, do have a required dormancy period. Even though there is enough soil moisture, these seeds require scarification and do not germinate until the next season. However, seeds of a variety of tea-leaved willow in the Rocky Mountains both ripen and germinate in July. In contrast, short-fruited willow seeds ripen in August when the habitat has dried, preventing germination until the next season. Other plants produce immature seeds that require winter dormancy, not fully ripening until the following season.

High temperatures are required for successful germination. The lower threshold for most plants is 40° F (5° C), but alpine sorrel requires at least 60° F (15° C). A diurnal cycle between cold and warm is usually necessary. Soil warming from solar radiation is important, but if surface soil becomes too hot or dry, seedlings die. Because roots must reach a water source before the surface soil dries, growth during the first year of most seedlings is mainly root development.

Because flowering, setting seed, germination, and survival of seedlings are difficult, most plants also reproduce by vegetative means. Such clonal development takes many forms, including rhizomes, stolons, off-setting new rosettes, growing adventitious roots from buried stems, and growth of bulbils or plantlets. The three most dominant alpine families (sedge, grass, and sunflower) produce obligatory clones, meaning that the new plants do not remain connected to the original plant but become totally separate individuals.

Plants live longer at higher elevations, up to 30–50 years for tall perennial forbs according to age estimates based on slow growth rates or DNA. Yellow stonecrop lives for five years, viviparous bistort for 26 years, and tussocks of kobresia for 200–250 years. Woody dwarf shrubs are also long lasting, 109 years for mountain cranberry. *Acantholimon diapensoides* cushions in the Pamirs are estimated to be 400 years old. Other slow-growing tap-rooted species live for 100–300 years. Some clones of carex sedges, cottongrass, alpine azalea, or willow can be thousands of years old. In New Zealand, *Raoulia* mats are hundreds of years old. The longevity of alpine plants makes the ecosystems especially susceptible to environmental damage because plants cannot quickly be replaced.

## Animal Adaptations

Animals must contend with reduced oxygen and sometimes strong winds. The reduction in oxygen content of the air is usually not a factor in limiting distribution of animals in the alpine. Yaks are permanent residents up to 19,500 ft (6,000 m) in the Himalayas. The guanaco of the Andes has more hemoglobin in its blood to help bind more oxygen, and many animals have hemoglobin with an increased affinity for oxygen. Llamas more efficiently extract more oxygen from thinner air. Although more red blood cells to carry more oxygen is not the norm for most alpine birds or mammals, it is a common acclimatizing response in lowland species, like humans, when traveling to higher elevations. Several high-elevation animals have higher heart and respiratory rates to circulate oxygen more quickly. Humans who live at high elevations develop larger lungs and therefore more gas-exchanging alveoli.

Unlike the Arctic, alpine tundra supports several burrowing or digging animals. Small animals usually require a deep insulating cover of snow over their winter homes. Meadow voles and water voles in alpine environments are restricted to lee slopes and depressions where snow accumulates. Red-backed voles and heather voles find protection under krummholz, while deer mice, pika, and mountain voles prefer rocky areas. Although low-lying areas are important for snow accumulation and insulation, those same areas are subject to flooding during snowmelt. Water voles live in especially wet meadows and can even swim, but their burrows must be elevated above flooding. Pocket gophers live in a narrow habitat zone between too wet with flooding and too dry with insufficient snow cover. Soil from tunnels they dig beneath the snow is left on the surface, exposed as long, sinuous piles of dirt when the snow melts.

Ungulates in mid-latitude alpine environments tend to congregate in small groups. Many are sure-footed and can maintain their balance on sheer rocky outcrops and boulders. Chamois and ibex in Europe, thar in the Himalayas, llama in the Andes, and mountain goat and bighorn sheep in North America, escape danger by jumping and climbing steep cliffs. Large front hooves, long toes, or hooves with gripping action are adaptations to their rocky habitat. Rocky areas, common in the alpine, are preferred nesting sites for many small mammals because they provide protection from powerful digging predators such as brown bear and badger. Marmots, pikas, and golden-mantled ground squirrels select large rocks to burrow beneath.

Breeding in alpine birds usually coincides with the peak of insect populations during the short summer. Flies, stoneflies, beetles, moths, butterflies, and the parasitic sawflies and bees are abundant. Arthropods are generally small, enabling them to hide easily in loose rock. Their dark color absorbs more sunlight, which is important in the cool air. Many have small wings or are flightless, a response to strong winds. Others can withstand ice in their tissues or undergo supercooling so their body tissue does not freeze.

## Treeline

Treeline generally occurs at lower elevation with increasing latitude, from more than 12,000 ft (3,600 m) in the Himalayas down to sea level in the Arctic, but more factors are involved than just latitude (see Table 3.1). Treeline is not a strictly defined line but a zone because microhabitats caused by late snowmelt, cold air drainage, sunny ridges, avalanches, cold or wet soil, and rocky slopes all affect growth of trees. Treeline is higher in both continental areas and larger mountain masses such as the Rocky Mountains and Himalayas because they retain more heat, making summers warmer. A combination of winter snow, summer drought, and more bare rock in the Sierra Nevada depresses treeline. Treeline is also lower in maritime areas, caused either by cooler temperatures due to ocean influence or by heavier snowfall and shorter growing seasons. The Pyrenees, for example, with both a smaller area and the influence of moist maritime airmasses, have a lower treeline than the Rocky Mountains at similar latitudes. For the same reasons, perhaps exacerbated by drought, treeline in the Southern Hemisphere is usually lower than that of similar latitudes in the Northern Hemisphere.

The woody plants at treeline are predominantly evergreen because of the growth advantage the evergreen habit affords during short cool summers. Pine is

**Table 3.1 Mid-Latitude Treeline According to Latitude**

| | | TREELINE | | |
|---|---|---|---|---|
| MOUNTAIN RANGE | LATITUDE | FEET | METERS | MAIN SPECIES |
| Northern Fennoscandia, Norway | 71° N | sea level | sea level | White birch<br>Scotch pine<br>Norway spruce |
| Denali, Alaska, United States | 64° N | 2,500 | 750 | White spruce<br>Black spruce |
| Central Fennoscandia, Norway | 61° N | 3,600 | 1,100 | Scotch pine<br>Norway spruce |
| Grampians, Scotland | 57° N | 1,600 | 500 | Scotch pine |
| Northern Rocky Mountains, Alberta, Canada | 50° N | 7,200 | 2,200 | Engelmann spruce<br>Subalpine fir<br>Alpine larch<br>Western hemlock |
| Mt. Olympus, Washington, United States | 47° N | 4,500 | 1,400 | Mountain hemlock<br>Subalpine fir<br>Alaska cedar |
| Central Alps, Switzerland | 46° N | 6,500 | 2,000 | European larch<br>Stone pine<br>Prostate pine<br>Norway spruce<br>Mountain pine |
| Carpathians, Romania | 46° N | 5,600 | 1,700 | Prostrate pine<br>Norway spruce |
| Cascade Range, United States | 45° N | 5,000 | 1,500 | Whitebark pine<br>Subalpine fir<br>Engelmann spruce<br>Mountain hemlock<br>Alpine larch |
| Mt. Washington, New Hampshire, United States | 44° N | 5,000 | 1,500 | Balsam fir<br>Black spruce<br>Paper birch |
| Central Rocky Mountains, United States | 44° N | 9,500 | 2,900 | Engelmann spruce<br>Subalpine fir<br>Whitebark pine<br>Limber pine |
| Southern Alps, New Zealand | 44° S | 3,300 | 1,000 | Southern beech |
| Pyrenees, Spain and France | 42° N | 7,500 | 2,300 | White fir<br>European beech |
| Tien Shan, Kyrgyzstan | 42° N | 9,500 | 2,900 | Central Asian spruce |

*(Continued)*

**Table 3.1** (*Continued*)

| MOUNTAIN RANGE | LATITUDE | TREELINE | | MAIN SPECIES |
| | | FEET | METERS | |
| --- | --- | --- | --- | --- |
| Southern Rocky Mountains, United States | 40° N | 11,500 | 3,500 | Engelmann spruce Subalpine fir |
| Andean Steppe, Argentina | 40° S | 5,400 | 1,650 | Southern beech |
| Sierra Nevada, California, United States | 38° N | 10,800 | 3,300 | Whitebark pine Mountain hemlock Sierra lodgepole pine Foxtail pine |
| Himalaya Mountains and Tibet | 30° N | 12,000 | 3,600 | Himalayan fir Himalayan pine Western Himalayan spruce Drooping juniper |
| Lesotho Plateau, South Africa | 29° S | 7,500 | 2,300 | Protea |

the dominant family, with species of pine, spruce, and fir most prevalent, but hemlock and juniper are also present. Larch (also called tamarack), the only deciduous conifer, has species in both western North America and in Eurasia. Hemlock forms treeline in the Cascade and Coastal ranges of North America. Broadleaf deciduous species are also common. Treeline in Fennoscandia is dominated by birch. Broadleaf trees are usually found where competition from conifers is absent, such as on disturbed areas in avalanche tracks. Their flexible stems and ability to regenerate from roots to form clones makes them ideally suited to the annual mass movement of snow. Alder is found on moist slopes where avalanches occur in both the Alps and British Columbia. Beech is dominant in central and southern Europe and the Caucasus. Rhododendrons and junipers form treeline in Nepal. Rhododendrons are trees in the Himalayas, not the shrubs familiar to Americans and Europeans.

Because of the physical and climatic continuity of North America and Eurasia, especially at the height of the Pleistocene, many plants in Northern Hemisphere mountains are similar throughout the mid-latitudes. Treeline is no exception, and different species of common genera may occur in different regions. A relationship between Clark's Nutcracker and whitebark pine in the Cascades, Sierra Nevada, and the Rockies is matched by similar relationships in other areas, such as Spotted Nutcracker and Swiss stone pine in the southern European mountains, from the Alps to the Caucasus, and Japanese Nutcracker and Siberian dwarf pine in eastern Asia. By feeding on and burying the pine seeds, the birds unwittingly disperse the trees.

Conspicuous in Northern Hemisphere mid-latitude mountains is a belt of stunted, crooked trees called krummholz (see Figure 3.2). Engelmann spruce, a common treeline species, is a good example. When tall individuals grow densely packed in the forest, the lower limbs die due to shade. However, isolated trees

**Figure 3.2** Under severe windy and icy conditions, trees become stunted, forming krummholz. *(Photo by author.)*

maintain healthy lower branches that may then form adventitious roots where they touch the ground. Over time, a small grove of cloned trees develops near treeline. In slightly harsher conditions of cold and wind, the growthform changes. A strong prevailing wind will either break branches on the windward side of the tree or bend them toward the lee side, forming "flag" trees. With even more extreme conditions, the plant becomes a shorter tangle of stunted limbs. Any shoot that grows above the protective winter snow cover is naturally pruned by cold, dry winds. Weight of heavy snow may also break exposed branches. The prostrate krummholz vegetatively reproduces by rooting downwind where it is protected by the snow it accumulates, while the windward side dies. A few trees, such as dwarf mountain pine in the European Alps, Siberian dwarf pine in eastern Asia and Japan, and some junipers, are genetically controlled forms that grow prostrate or dwarfed even under nonalpine conditions.

Treelines in Southern Hemisphere mid-latitudes are more floristically diverse than those of the Northern Hemisphere due to isolation and lack of convenient migration routes. Most trees and shrubs are broadleaf evergreens. The most prominent is the Southern beech, distributed in South America as well as Australia, New Zealand, and as far east as New Guinea, even though it is not adapted to long-distance dispersal. Deciduous species of southern beech occur in the drier mountains of Patagonia and New Zealand. In southern Africa, treeline in the Drakensburg Mountains is a grassland and scrubland because of a dry climate and frequent fires. The highest tree is a *Protea*.

## Habitats and Plant Communities

Alpine plants are usually limited to a specific environment, but plant communities change from one to another over short distances, perhaps just a few feet, because of the small-scale mosaic pattern of habitats (see Figure 3.3). A boulder field, also called felsenmeer—literally a sea of rocks—is a large accumulation of angular rocks shattered by frost action. Rocks completely cover the surface, which may be either fairly level or on a gentle slope. Boulder field communities are largely limited to crustose and foliose lichens on the jumble of large rocks. Because of the consistent growth rate of some lichens—such as map lichen, which requires 1,000 years to reach a diameter of 0.4 in (1 cm)—boulder fields can be dated according to when they were last covered by glacial ice. A few vascular plants may grow in pockets between rocks where soil collects, snow accumulates, and the rocks both radiate heat and provide wind protection. Pikas and marmots take shelter in boulder fields, especially if close to a meadow for food.

Establishment of plant life on unstable talus and scree rock accumulations at the base of cliffs and steep slopes is difficult. On talus slopes, the smallest rocks are closest to the top of the cliff, while the largest are carried by gravity to the base. The sharp, angular rocks of talus, a product of frost action, are as large as 6 in (15 cm) in diameter. Scree refers to steep slopes of smaller rocks, down to gravel size. Although apparently dry, damp soil may be found just 1 in (2.5 cm) beneath the scree surface. Mountain avens, a circumpolar plant occupying the same habitat in both arctic and alpine environments, is the best pioneer plant to stabilize scree. With thick evergreen leaves covered by shaggy hairs on the underside and an elaborate root system intertwined in scree, it survives downslope movement on windy slopes where other plants would be ripped apart. Avens has nitrogen-fixing nodules on its roots and enriches the soil with nitrogen. As branches accumulate dirt and debris, humus piles develop beneath its leaves. Other plants on scree include Easter-daisies, alpine milkvetch, buckwheat, and phlox. Roots are either a mesh of shallow rhizomes or big taproots which provide security on the slope.

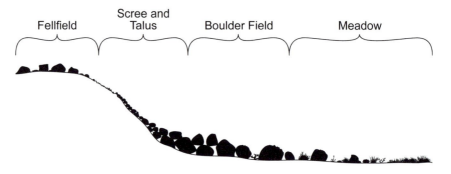

**Figure 3.3** Plant growthforms and species change across environmental gradients. *(Illustration by Jeff Dixon.)*

With 35–50% rock, fellfields are flat areas usually remaining bare of snow due to wind. Soil development between the rocks is poor because wind carries away fine particles, leaving only shallow coarse gravels. Because of rapid drainage, summers with little snowmelt, and wind exposure, fellfields are dry environments where plants are exposed to extremes. Major growthforms are ground-hugging cushions, mats, succulents, and rosettes, which may also be hairy or have waxy cuticles. Typical plants, which bloom early in spring before any available moisture is gone, include moss campion, alpine phlox, and sandwort. Few mammals live in fellfields because forage is scarce.

Tussock-forming grasses, sedges, and forbs about 8 in (20 cm) tall grow in meadows and bloom in summer because soils are moist. Lichens and mosses cover the ground between higher plants. Sedges are dominant but both sedges and grasses are present and both extend their roots 1 ft (0.3 m) or more into the deep soils, the best on the alpine tundra. Because leafy forbs are abundant, meadows are sometimes referred to as herb-fields.

Snowbeds develop where snow accumulates in the lee of wind obstructions, such as plants or rocks, or in nivation hollows that develop where snow accumulates on a slight slope (see Figure 3.4). The weight of the snow pushes saturated soil downslope into a low ridge, creating a shallow bowl or depression that then accumulates more snow and intensifies the process. The pattern of late-melting snow causes concentric rings of plants to develop, reflecting the time of snowmelt and length of the growing season. Plants in winter are protected from intense cold and temperature changes by deep snow. Temperatures in snowbeds are warmer compared with exposed fellfields. Snow also provides protection from drying winds. Dust and debris particles that collect on the snow surface add to soil development when the snow melts. The pink tinge of old snow is caused by green algae encased in a reddish coating that survives at melting snow temperatures of 32° F (0° C). The snow surface attracts flies, spiders, mites, and springtails where they feed on

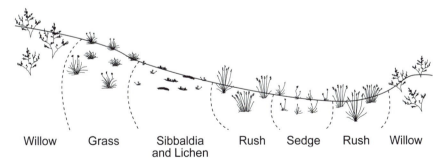

Willow     Grass     Sibbaldia     Rush     Sedge     Rush     Willow
                     and Lichen

**Figure 3.4** Snowbed communities typically have rings of different plants, each growing according to date of snowmelt and length of growing season. Snow in the center may not melt all summer. (*Illustration by Jeff Dixon. Adapted from Zwinger and Willard 1972.*)

pollen and spores. Insects in turn attract birds. However, snowbeds also have detrimental effects, such as sheer weight of the snow, lower summer soil temperatures, too much water, solifluction, and a shortened growing season. Snowbed indicator plants are widespread. Snow liverwort, dimpled lichen, and flake lichen all serve to bind particles and stabilize soil. Scandinavians use the presence of creeping sibbaldia, a snowbed plant, as an indicator of soggy soils that would make a poor roadbed.

Animal-disturbed communities develop most commonly in meadows and snowbeds where soil is both deep and well drained enough for digging. In North America, pocket gophers are most often the culprits. The gophers' bare dirt mounds are subject to needle ice, which causes further disturbance. The sites often revert to fellfield vegetation with cushion plants. Water and wind erosion may further degrade the soil, but soil disturbance can be beneficial because it increases porosity, allowing rain and snowmelt to soak in. Meadow voles take over abandoned gopher holes and tunnels. They eat grasses and sedges in wet meadows and undergo three- to four-year population peaks. Active in winter beneath the snow, they can destroy cushion plants if their numbers are high enough.

Behind solifluction terraces on permafrost, around lake edges, or beneath snowbanks are wet marsh communities that remain green all summer because of the abundant water supply. These areas, with mosses, lichens, sedges, and low willows, most closely resemble arctic tundra. They are high in organic matter and peat, which insulates and preserves the permafrost beneath. Soil is thick, black, and mucky on top with many roots. Below the surface layer, the soil is colored by blue-gray clay or red-yellow iron stains. With humic acids and no oxygen, the soil is acidic. A common plant is marsh marigold.

Heath communities are widespread in the northern Rockies in both the United States and Canada and are also found in the Olympics, Cascades, Sierra Nevada, and Europe (see Figure 3.5). Because they prefer cool cloudy weather, they are rare in the southern Rockies where it is too dry and sunny. All heath plants are members of Old World genera and grow in acidic and well-drained but moist soils. Protective winter snow melts early in summer. Blueberries, bilberries, bell heather, Labrador tea, alpine azaleas, and rhododendron are common. Plants, many of which are evergreen, are low growing with small leathery (sclerophyllous) leaves. Their thick cover and short stature prevents establishment of most plants other than Iceland lichen.

## Pleistocene Migrations and Similarities of Floras

Of the approximately 1,000 arctic tundra species, 500 of them are also found in Northern Hemisphere mid-latitude alpine. A few genera (bluegrass, draba, sandwort, cinquefoil) are also in the tropical alpine, and spiked trisetum grass can be found all the way to Antarctica. Mount Washington in New Hampshire has 75 alpine species, almost all of which are arctic. In the Sierra Nevada with more than

**Plate I.** Landscapes affected by continental glaciation, shown here in northern Manitoba, Canada, are usually gently rolling with rocky outcrops and erratic streams. *(Photo by author.)*

**Plate II.** Mountains subjected to alpine glaciation, viewed here from the summit of Mount Whitney in the Sierra Nevada, change from rounded uplands to deep U-shaped valleys between sharp-edged ridges. *(Photo by author.)*

**Plate III.** Several animals in tundra environments seasonally change color. White-feathered ptarmigan are hard to see in a snowy winter landscape, and mottled plumage camouflages the birds against the willows and rocks in summer. *(Courtesy of Shutterstock. Copyright: Mark Yarchoan and Jack Cronkhite.)*

**Plate IV.** Muskoxen are one of the few animals that survive the harsh environment of the High Arctic. *(Courtesy of the U.S. Fish and Wildlife Service.)*

**Plate V.** Cottongrass sedge is typical of much of the Arctic tundra in Canada. *(Courtesy of the U.S. Fish and Wildlife Service.)*

**Plate VI.** Mountain azalea is a common dwarf shrub in both Arctic and mid-latitude alpine tundra. *(Photo by author.)*

**Plate VII.** Because they depend on sea ice for breeding, Emperor Penguins are threatened by global warming. *(Courtesy of Shutterstock. Copyright: Jan Martin Will.)*

**Plate VIII.** Bristlecone pine, here on Mount Moriah in the Snake Range of Nevada, lives for more than 3,000 years. *(Photo by author.)*

**Plate IX.** Alpine environments of the Rocky Mountains have a variety of wildflowers. *(Photos by author.)*

**Plate X.** Mount Washington in the White Mountains of New Hampshire is famous for having some of the worst weather in the United States. *(Copyright © Matthew Gilbertson. Used by permission.)*

**Plate XI.** Sheer, rugged peaks rise above alpine meadows in the Dolomites region of the European Alps. *(Photo by author.)*

**Plate XII.** The snow leopard is a rarely seen resident in the alpine tundra of the Himalayas. *(Courtesy of Shutterstock. Copyright: Joseph Gareri.)*

**Plate XIII.** Large azorella cushion plants grow in the Altiplano at 14,000 ft (4,250 m) north of Arequipa, Peru. Nevado Huarancante, rising to 16,500 ft (5,000 m), is the mountain in the background. *(Copyright © Dr. James S. Kus.)*

**Plate XIV.** Puya (*Puya hamata*) is a giant rosette common in both the páramo and puna in the Andes. *(Courtesy of the author and Susan L. Woodward.)*

**Plate XV.** Two major giant rosettes cover this slope on Mount Kenya, giant lobelias in the foreground and giant groundsels in the background. *(Courtesy of Rainer W. Bussmann, Missouri Botanical Garden.)*

**Plate XVI.** Silversword grows on the cinders of Haleakala crater on Maui, Hawaii. *(Photo by author.)*

**Figure 3.5** Heath communities, such as in Öztal, Austria, in the European Alps, are characteristic of cloudy conditions and moderate temperatures. *(Photo by author.)*

600 species, only 20% are arctic; most others are related to desert lowlands, the Rocky Mountains, or the Cascades. The degree of relationship between the Rocky Mountain flora and that of the Arctic depends on location, with areas farther north having more arctic elements. In the Wrangel Mountains in Alaska, 70% of the plants are arctic, while in the Sangre de Cristo Range in the southern Rockies, only 32% are. Eurasian mountains also show a relationship with arctic plants according to latitude. In Fennoscandia, 63% of species are circumpolar. The Swiss Alps and the Altai Range have 35% and 40% arctic species, respectively.

Central Asian Mountains (Altai, Himalayas, and Karakoram) have large and distinctive floras because they were not completely covered by glaciers, and thus have had a longer time for speciation to occur. The Himalayas are believed to have been a Pleistocene refuge, and many alpine plants may have originated there. Of the more than 1,000 alpine species in the Himalayas, only an estimated 25–30% are also found in the Arctic.

## Arctic to Rockies to Andes

The Pleistocene era probably contributed more to the current distribution of arctic and alpine plants in North America than any other factor. Many species in both biomes are identical, and many more are closely related members of the same genera. Many plants originated in the Arctic and migrated to the Rocky Mountains, while others originated in mid-latitudes and followed the retreat of cold climates northward at the close of the Pleistocene. These migrations were facilitated by the north-south orientation of mountain ranges and their continuation into South America. Even though the mountain chains are not continuous, peaks were close enough together to allow the passage of plants.

With the onset of the Pleistocene, temperatures in the Rocky Mountains and other North American ranges decreased and glaciers formed on many peaks and ranges. The arctic tundra vegetation spread southward along the axes of the ranges, taking advantage of the cooler climate. Even though no glaciation occurred south of the San Juan Mountains in southwestern Colorado, herbaceous tundra plants were able to find suitable habitats all along the peaks in the Rocky and Sierra-Cascade mountain chains all the way to the Andes and even Tierra del Fuego at the southern tip of South America. At least 19 families of seed plants in the Colorado alpine are also represented in the Andean páramo above 11,500 ft (3,500 m), and a minimum of 20 genera are shared by both regions. However, at least 22 additional families in the páramo are not represented in the Rockies, and few to none of the same species occur in both the Rockies and the páramo. The greater variety of plant families in the páramo and their absence in the mid-latitudes seem to indicate that the general trend of migration was to the south, with little northward movement. A partial explanation may be that in the late Pleistocene, parts of the Andes were still being built by volcanic activity and suitable habitats were destroyed, blocking northward migration. (See Chapter 4 for discussion of Andean páramo.)

Mid-latitude plants were also adapting to an alpine climate. Depending on the time period, these adaptations may have taken place as the general land surface was uplifted from lower, subalpine, or even desert elevations, or in response to the cool, periglacial climate. With each glacial retreat, plants—both the original arctic species and the newly derived mid-latitude species—followed the cooler climates north, sometimes migrating from peak to peak. Arctic species most likely advanced and retreated with each change in climate. Those species originating in the Rockies or in the Sierra-Cascade ranges may have adapted to the alpine climate at any time during the Pleistocene, some possibly in the last stages of the era, giving them less time and hence opportunity to migrate north and become established in the Arctic. This may partially explain why more arctic species grow in the mid-latitudes than vice versa.

## North-South Migrations in Europe

The major difference affecting migration in Europe versus the Americas is the trend of most of the mountain ranges. Unlike the north-south corridors offered by the Rocky Mountains, the east-west ranges of Europe served as barriers to both northward and southward migration. The flora of the European Alps is more closely related to the mountains in central Asia than it is to the flora of Fennoscandia. Similarities in genera and species do exist, however, between the Alps and the Rocky Mountains. One possible explanation is migrations from North America over the Bering Strait land bridge and subsequently west to the Alps—a very long journey. Some species, such as mountain avens, alpine sorrel, moss campion, alpine bluegrass, and nodding saxifrage, are common to both the arctic and to the alpine areas of Europe.

During the Pleistocene, almost the entire British Isles were covered with ice; tundra climate and vegetation developed at the edges of the ice sheet. As the ice retreated and climate warmed, temperate vegetation from Europe invaded, pushing the alpine flora to the highlands. Ample evidence survives in bogs to show that alpine vegetation was formerly much more widespread. Remains of alpine plants are found buried deep beneath peat that has accumulated since the Pleistocene.

## Adaptations of Lowland Species

Plants from some types of lowland environments may have rather easily adapted to alpine conditions as mountain ranges rose. Some nonmountain plants are early spring bloomers that die back during the hottest part of summer. Hot steppes or deserts are more likely to have species preadapted to life at high elevations because many of their plants bloom in spring. The fact that Indian paintbrush is found in both southwestern deserts and in the alpine tundra of the Rocky Mountains is a good example of the similarity between desert and high-elevation floras and environments. It might be expected, too, that boreal forest plants, adapted to cold temperatures, would also become members of the alpine flora, but they live in the shade of the forest and are not exposed to intense light. In many Asian mountains, the krummholz zone is poor in flora and many bald peaks occur because forest plants failed to adapt to the alpine zone.

In contrast to slow uplift, a relatively rapid uplift would eliminate the local nonmountain flora because of the lack of evolutionary time, thereby leaving the alpine tundra habitat open to colonization by any appropriate immigrating alpine species, or in their absence, leaving it species poor. In any case, the size and composition of the alpine flora depend on several factors other than the rate of uplift, including age of the mountain, proximity of nearest mountains, richness of flora on adjacent mountains, prevailing winds, and bird migrations.

## Regional Expressions of Mid-Latitude Alpine Tundra

### North America

Although not exceptionally high, the northern Appalachians in eastern North America have alpine areas (see Figure 3.6). Mountains in Labrador and on eastern

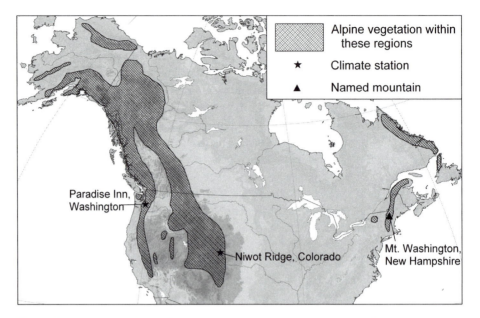

**Figure 3.6** Alpine tundra in North America is predominantly in the higher mountains in the West. *(Map by Bernd Kuennecke.)*

Ellesmere and Axel Heiberg Islands form the eastern fringe of the Canadian Shield. These mountains can be considered both arctic and alpine. Most North American mountains high enough for alpine environments, however, are in the western part of the continent, where two main cordilleras are dominant. Although not always directly on the coast but still with maritime influence, a coastal region extends from the Alaska Range south to the Cascades and Sierra Nevada in California. Many peaks are more than 14,000 ft (4,300 m). Denali (formerly called Mount McKinley) in southwestern Alaska, at 20,298 ft (6,187 m), is the highest peak in North America. The Cascades are still-active volcanoes up to 14,000 ft (4,300 m) high that sit atop a lava plateau at 5,000 ft (1,500 m). The Sierra Nevada is a recently uplifted fault block mountain reaching elevations greater than 12,000 ft (3,600 m). Except for the Olympic Range in Washington, most mountains directly on the coast in the United States are lower, and few rise above treeline.

In the interior, an almost continuous stretch of mountains extends from the Brooks Range in central Alaska south through the Rocky Mountains in western Canada and United States to northern New Mexico. The Rockies are geologically complex, with many peaks above 14,000 ft (4,300 m). The Great Basin, with its series of north-south-trending fault block mountains, lies between the western ranges and the Rocky Mountains. Some summits rise more than 13,000 ft (4,000 m), and even though relatively isolated, most of their flora resembles that of the Rocky Mountains. The volcanic San Francisco Peaks complex in northern Arizona is also an isolated mountain range between the Sierra Nevada and the Rockies.

All mountain ranges with alpine environments have been glaciated. Coastal mountains in Alaska still have extensive valley glaciers, as do the Olympic Mountains and several peaks in the Cascades, such as Mount Rainier and Mount Shasta. The drier Sierra Nevada and interior ranges have only small glacial remnants.

Alpine tundra in North America is geographically widespread, each region or range having a unique environment. They have much in common but also many differences. All have low air and soil temperatures, high winds, intense solar radiation, short growing season, and sometimes heavy snow. Geology and substrate vary and include all kinds of rocks. Northern mountain alpine regions are similar to arctic tundra where the ecosystems merge. Coastal mountains have frequent clouds and fog, and consequently less solar radiation. Continental mountains have fewer clouds and more solar radiation, but also intense summer thunderstorms. Permafrost is most common in northern mountains. Widespread circumpolar or arctic-alpine distributions are common, as are areas rich in endemics evolved from surrounding lowlands. Limitations imposed by the alpine environment restrict the number of species. The Colorado alpine environment supports about 300 species of plants, the White Mountains in California have about 150, and east-coast mountains have only about 100.

Because of prevailing westerly winds, the west coast and western sides of mountains receive more precipitation. Most precipitation in the Sierra Nevada and Cascades falls in the winter, with limited summer thunderstorms. The Cascades receive heavy winter snow, while summers in the Sierra Nevada are especially dry. All the interior mountain ranges, including those in the Great Basin, are susceptible to intense thunderstorms in the summer because of continental climate and surface heating in surrounding valleys. On the east slope of the Rockies, especially in Colorado, early fall snow is very wet and heavy because of moist air circulating upslope from the Gulf of Mexico. While the same source of moisture may cause an occasional light summer snowfall in the southern Rockies, it will more often generate thunderstorms. Summer snowfalls can provide much needed water to alpine vegetation. In contrast, winter snow is dry, icy, and associated with strong winds. The White Mountains in New Hampshire are not only in the path of cyclonic storms from the west but also get maritime influence from the Atlantic Ocean. Consequently, they are wet.

South of 60° N, permafrost is discontinuous and influenced by slope, aspect, and drainage. Permafrost plays only a small direct role in alpine vegetation because most plant roots are shallow and not restricted by the active layer, but soil disruption from frost action has indirect effects. Frost-heaving can affect soils in Colorado to a depth of 12 in (30 cm).

***Origins and plant communities.*** North American alpine regions have complex floras with a mixture of arctic-alpine, circumpolar, and lowland elements, with many endemic genera and species. Each site has a complicated geologic history. During the Cretaceous, North America continued to move northward into cooler latitudes, eliminating frost-sensitive species. Circumpolar species migrated across the Bering Strait land bridge both to and from Asia. Mountain building and glaciation isolated areas, and Pleistocene glaciation pushed arctic species south. During postglacial warming, arctic species and adapted lowland species were forced to retreat upward to the alpine. Northern parts of both the Sierra Nevada-Cascade and the Rocky Mountains are dominated by circumpolar and arctic-alpine elements, which become less important farther south. Lowlands like the Columbia River Gorge, currently barriers to alpine plants, were colder and passable during the Pleistocene. Mountains in the Great Basin are isolated peaks surrounded by desert shrubland. As such, they have fewer species, both because of isolation and small summits with limited alpine habitat. Rocky Mountain species extend west across the Great Basin, perhaps because the Rockies are older and served as a biotic source area for a long time period. Species from the Sierra Nevada, which is younger than both the Rockies and Great Basin mountains, are not found far east into the Great Basin.

Herbaceous vegetation is mostly perennial and includes cushion plants, rosettes, grasses, and sedges. Annuals are a minor component of arctic and alpine flora but are more commonly found in alpine situations. Some 55 annual species occur in alpine North America, 47 in the dry summer Sierra Nevada in California, 10 in the Rocky Mountains, and only two in New England. At least 10 annuals have both an arctic and alpine distribution. Based on exposure, snow cover, and moisture, a gradient from steep rocky slopes to the valley bottom forms a typical vegetation profile (see Figure 3.3). Woody cushion plants such as moss campion and purple saxifrage occupy the steep rocky fellfields exposed to wind and generally are snow free. Drier meadows have a mountain avens turf, while moist areas farther down the slope support hairgrass meadows. Where water accumulates in bogs sedges dominate, with willow thickets at streamside. Late-melting snowbanks may be barren except for lichens and moss-covered rocks because the growing season is too brief for vascular plants.

***North American alpine animals.*** Although many plant species are shared by both arctic and alpine areas, few mammals or birds are common to both. Alpine regions have more diversity of mammals. In the Cascade Range in Washington and Oregon, 47 mammals use the alpine environment, and 32 occupy Niwot Ridge in the Colorado Rockies. Smaller alpine areas have fewer. Animals can be divided into three groups: permanent residents, seasonal residents, and occasional visitors.

Rodents are plentiful, and although population cycles similar to those of lemmings in the Arctic do not occur, numbers do vary from year to year. Because small resident mammals need protective cover, habitats with rock piles, krummholz, and willow thickets have the highest species diversity. Burrowing animals other than pocket gopher are not common due to rocky soils, but they are important for soil enrichment. Their digging mixes and aerates the soil, adds nutrients from droppings, buries plants and humus, and distributes seeds. Depending on geographic location, some small resident mammals have narrowly defined habitats, while others are wide-ranging. Pikas, deer mice, and voles remain active beneath snow in winter. Mice and voles share communal nests to increase warmth. In contrast, marmots hibernate to escape the winter cold.

Erroneously called cony, which refers to a hoofed animal in the Himalayas, pikas are in the rabbit family but have small ears in response to the cold (see Figure 3.7). Widespread in North American alpine tundra, their territory can be recognized by the red-orange color of jewel lichens that grow on rocks where the animals perch. Pikas are solitary during the winter but mate in spring and raise the litter as a pair. They are restricted to rock piles or rock polygons for protection from predators but need access to meadows for food. The furry soles of their feet provide traction as they scurry across rocky terrain. Pikas remain active beneath the snow in winter, eating hay and seeds they harvested in summer. They cut meadow plants and place them on rocks in the sun to dry, thereby preventing mold from spoiling their food stores. Although grasses and sedges dominate, hay piles also include several species of forbs. Opportunistic feeders, pikas will take anything edible. Because their food has little nutrition, they must eat every hour. They also reingest fecal pellets to gain the most nutrients. They conserve moisture by expelling almost crystalline uric acids, which can be seen as a white coating on the rocks.

Marmots and woodchucks, or groundhogs, are found only in the Northern Hemisphere. They probably originated in the Himalayas, but migrated to North America where they now live in many environments, including grasslands and alpine tundra. Marmots grow up to 24 in (60 cm) long and weigh 15 lb (6.8 kg). All species spend summers storing fat for winter. They have thickly furred bodies, short ears and legs, and a bushy tail. Strong claws enable them to dig burrows in boulder fields close to meadows for food. Marmots will stay in their burrows when biting insects are especially active on windless days. They are not fussy eaters but prefer forbs over grasses and sedges. Rocks provide lookout stations for protection against predators, where marmot whistles alert the colony to potential danger.

Marmots are social. Some live in groups of 5–10 animals usually consisting of one breeding pair and its offspring, or one male with two to three females. Females breed at two to three years, bearing two to six pups depending on the species. Marmots live about five years, and pups remain together as a family group for two to three years before venturing out on their own. New males moving into a social group often kill any pups fathered by former resident males.

Marmots have both large and small enemies. They provide summer food for foxes, coyotes, wolves, bears, and eagles, and they are also hunted by humans. Mongolians and Native Americans use the meat and fur.

**Groundhog Day**

Marmots and groundhogs are true hibernators, lowering their body temperatures to slightly above freezing. The hibernation period varies with species and location, but may be one-half to two-thirds of the year. Even though in hibernation, marmots and groundhogs do wake up occasionally. Punxsutawney Phil, a groundhog in Pennsylvania, is famous for "predicting" the duration of winter based on whether or not he sees his shadow when he comes out of his burrow on February 2, Groundhog Day.

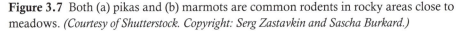

**Figure 3.7** Both (a) pikas and (b) marmots are common rodents in rocky areas close to meadows. *(Courtesy of Shutterstock. Copyright: Serg Zastavkin and Sascha Burkard.)*

Kobresia, hairgrass, and other alpine sedges and grasses provide nutritious forage for elk, deer, and mountain sheep. Some alpine areas are currently used for grazing of domestic stock—cattle and sheep. If overgrazed by domestic animals, the cropped tussocks cannot trap the snow, which is their main source of moisture on dry, windy ridges and slopes, and subsequently die.

Few large carnivores are found in alpine tundra. Brown (or grizzly) bears are now limited in distribution. Mountain lions and black bears are occasional visitors in the summer but migrate to lower elevations in winter. Coyotes venture onto the tundra in the summer to hunt rodents in wet meadows. Short-tailed weasels hunt in boulder fields, slipping their slender bodies into rodent burrows. Because prey is always available, weasels are active all year. Shrews feed on insects and other small invertebrates.

White-tailed Ptarmigan is one of the few birds to occur in both arctic and alpine regions. Only the male remains on alpine tundra all winter; the females winter in willow thickets at treeline. Triggered by longer daylight hours in early May, males occupy territories in which they are joined by the females. The birds usually pair for life even though they separate after breeding and the female raises the chicks alone. By necessity, the nest is a depression on the ground, usually hidden among plants. Chicks and eggs are preyed on by nutcrackers, falcons, and weasels, and also succumb to harsh weather conditions. Ptarmigan females often feign a broken wing to lead an intruder away from the nest. Although ptarmigan is the only permanent resident, several breeding birds that winter elsewhere are common to most of the North American mountains. The Water Pipit, found throughout the alpine zone in western United States, winters in South America, but from late June to August, it resides on the alpine tundra, eating a variety of insects. While some birds fly long distances to avoid winter cold, others, such as Rosy Finches, just move into lower elevations. Golden Eagle and Red-tailed Hawk are raptors frequently seen in the west (see Figure 3.8). Only the Red-tailed Hawk is common on the east coast. Other birds often seen in krummholz at the edge of western alpine zones are Stellar's Jays and Clark's Nutcrackers.

**Figure 3.8** Redtail Hawk is a frequently seen predator in alpine environments. *(Courtesy of Shutterstock. Copyright: Jemini Joseph.)*

As in the Arctic, cold temperatures restrict amphibians and reptiles, both of which are rare in alpine tundra. Insects are plentiful and include major pollinators, such as bumblebees and flies. Most prevalent are black flies, mosquitoes, craneflies, and many types of tiny flies. Butterflies are sometimes carried by wind from lower elevations. Grasshoppers, ladybugs, beetles, leaf hoppers, and spiders also live at these elevations.

*Coastal ranges.* Treeline species in the coastal ranges, dominated by mountain hemlock, subalpine fir, and Alaska cedar, are typical of the distinct maritime influence on the west coast. The alpine zone consists of various combinations of low shrubs and perennials. The northern alpine zones have more affinity to arctic tundra, while areas farther south are more similar to nearby alpine tundras. Windy, snowfree ridges facing south or west are sites for fellfields or boulder fields where rosettes, cushions, and prostrate shrubs with a ground cover of foliose lichens dominate. Moister sites on moderate slopes with intermediate snow depths support meadows of sedges and fescue grasses with small birch or willow shrubs. Alpine

**Olympic Mountains**

Although not exceptionally high—the summit of Mount Olympus is only 7,965 ft (2,428 m)—the Olympic Mountains are the wettest in the United States with long and active valley glaciers. Directly in the path of westerly winds from the warm North Pacific Current, the mountains have a mean annual precipitation of 200 in (5,000 mm). The lower slopes are densely covered in temperate rainforest. The northeastern slopes, however, are in a rain-shadow and receive only 20 in (500 mm) annually. Because of abundant moisture, alpine vegetation is dense and rich. Tall glacier lilies, grasses, paintbrushes, and louseworts grow in profusion in wet subalpine meadows.

bittercress and creeping sibbaldia dominate snow-beds and moist depressions, but an alpine turf of mountain hare sedge with many flowering forbs, especially chiming bells, may also develop. Drainage sites that get nutrients and water from snow-melt upslope support sphagnum hummocks with black crowberry and arctic willow. Fireweed, tufted saxifrage, and many more species grow on stabilized rocky slopes, but unstable scree slopes are sparsely populated. Watered areas with good drainage and protection from wind are covered by low carpets of mountain heather.

Rodents, including deer mice, meadow voles, and pikas, are the most common animals in the coastal ranges. The hoary marmot is widespread in Alaska and Canada, but except for the Olympic marmot, which is endemic to the Olympic Range, no marmots live in the coastal ranges in the United States. The Mazama pocket gopher is restricted to the coastal ranges and the Cascades. The mountain goat has been introduced to the Olympic Mountains. With no natural enemies to restrict populations, its sharp hooves cause increased soil erosion and alteration of plant communities. Dall's sheep is native to the Canadian and Alaskan mountains. Short-tailed weasel and dusky shrew are the most prevalent small carnivores. Brown bears are limited to the coastal ranges in Canada and Alaska.

Birds, such as White-tailed Ptarmigan, Horned Lark, Water Pipet, and Stellar's Jay are similar to those found in most North American alpine tundra. Golden Eagle and Red-tailed Hawk are commonly seen raptors.

*Cascade Range.* The Cascades extend from southern British Columbia to northern California. The western side of the range is maritime and wet, with rainfall equivalents of 100 in (250 mm) a year, while the lee side is drier and more continental, receiving 20 in (50 mm) or less of precipitation. Several large valley glaciers extend to elevations below the alpine zone, especially in the North Cascades and on Mount Rainier and Mount Shasta. Treeline varies with age of the volcanic peaks, but is about 5,000 ft (1,500 m). Trees are those typical of coastal mountains, but include whitebark pine and Engelmann spruce on the drier eastern side. Alpine communities are poorer than those in the Rockies due to substrate conditions. Soils, undeveloped because of recent volcanic activity, provide few nutrients, have little humus, do not retain water, and can be abrasive to plant shoots and roots. Much of the alpine zone on Mount Hood is covered with loose pumice. Both fell-field and snowbed plants are sparse and small because volcanic soils are unable to retain water.

Good volcanic soils support black crowberry, which helps stabilize steep slopes. Other typical plants include cliff Indian paintbrush and Lemmon's rockcress. Tolmie saxifrage and partridge foot, a tiny shrub, often grow together on steep scree slopes, helping to stabilize the loose volcanic soil. Windy fellfield communities have few species, mostly grasses, sedges, and rushes, with no annuals, although prostrate pussypaws, Lyall lupines, and alpine asters do occur. Bistort can be found in more protected microhabitats. Newberry knotweed is characteristic of open pumice areas in the Cascades where it has no competition.

The Cascades in northern Washington, especially Mount Baker and Glacier Peak, are geologically complex and heavily glaciated, with a varied environment due to different kinds of rocks. A strong floristic affinity with the Arctic and mountains to the north and east exists, with few endemics. Deep snowbeds have sedge mats with hairgrass and tiny partridge foot shrubs. Herb communities on moist southern exposures are dominated by lupine. Dwarf-shrub heath communities are dominated by kinnikinnik, western moss heather, or mountain heather. Associated species vary with location and conditions. Kobresia sedges form a dense cover in dry meadow communities. Boulder fields are characterized by patterned ground and high floristic diversity.

Rodents in the Cascades have habitat preferences. Mazama pocket gophers are common in grasslands, herb-fields, and wet meadows. Rock piles adjacent to meadows, providing both protection and nearby foraging areas, are home to pikas and yellow-bellied marmots. Water voles occupy willow thickets, and heather voles take cover in krummholz. Dusky shrews are restricted to wet meadows, whereas deer mice are found in rock piles, herb-fields, meadows, willow thickets, and krummholz. Short-tailed weasels prey on rodents. Pacific winds often bring storms that cover everything with ice for days or weeks, making foraging difficult. Mountain goats have been introduced. Ptarmigan is present, and summer bird populations are the same as those in the Rocky Mountains, including Rock Wren and Clark's Nutcracker.

*Sierra Nevada.* A tilted fault-block mountain primarily composed of granite, the Sierra Nevada range extends 400 mi (650 km) from northern to southern California. It contains the highest peak in the contiguous United States, Mount Whitney at 14,496 ft (4,418 m). Precipitation falls in winter, with potential for high snowfall, although annual totals are variable. The sandy soil is dry, with near-desert conditions throughout most of the summer. Summer drought is responsible for the absence of many arctic and alpine plants more common in the Rockies. Timberline

**Mount Rainier**

Dormant but still warm at its summit crater, Mount Rainier's volcanic cone has been extensively eroded by radiating valleys filled with alpine glaciers. Mount Rainier receives record snowfalls, and glaciers cover a significant part of the mountain. At Paradise Inn (elevation 5,550 ft or 1,692 m), slightly above treeline, the average annual precipitation is 116 in (2,945 mm), with annual snowfall of 50 ft (15 m). In 1971–72 the station received the world record of snowfall in a single season, 90 feet (27.5 m). Glaciers, however, have been melting over the past several years.

species reflect the difference between the wetter western slope and the drier eastern slope. Mountain hemlock is a major timberline tree on the west. On the eastern slope, foxtail pine resembles bristlecone pine, a treeline species in the Great Basin, in its age and gnarled appearance, and whitebark pine forms flagged krummholz.

The Sierra has the richest alpine flora in North America, but less than 20% of the species are of arctic origin, compared with 50% in the Central Rockies. Another 17% are endemic and related to genera at lower elevations in California or the Great Basin. A few large genera, such as milkvetch, lupine, cinquefoil, penstemon, and buckwheat, account for many of the species and subspecies. High plateaus not covered by ice during the Pleistocene (nunataks) have the richest flora because they served as refugia. Almost all major Rocky Mountain alpine communities are present on the unglaciated plateaus.

The Sierra flora has a relatively large component of annual species, predominantly on dry, sandy sites such as south-facing slopes. Several plant communities are based mostly on moisture content but to some extent on substrate (see Figure 3.9). Wet marble areas support a dense meadow of short-hair reedgrass and other grasses and sedges, with some prostrate shrubs of dwarf bilberry. Slightly drier upland sites with more exposed soil have a saxifrage-pussytoes cushion plant community. Short-grass sedge and Nutall's sandwort grow with timberline sagebrush shrubs in dry meadows. Widely spaced rosette and cushion plants dominated by oval-leaved buckwheat and endemic clubmoss ivesia grow in alpine gravels. Plant

**Figure 3.9** Granite fellfields in the Sierra Nevada are too dry to support many plants. *(Photo by author.)*

cover on bare granite is sparse, primarily phlox, Davidson's penstemon, and ivesia. Very arid sites are bare except for pussypaws and dwarf knotweed. Carson Pass, where andesite lava flows replace granite, is the northern limit for many alpine species. The andesite supports few plants except for alpine sorrel where snow melts late. Snowfree andesite sites have plants more closely related to cold deserts at the base of the Sierra. Common plants are bottlebrush squirreltail, Great Basin violet, and prickly gilia.

In addition to widespread rodents, such as pika, deer mice, and heather vole, the Sierra is home to the endemic alpine chipmunk and Sierra pocket gopher. Yellow-bellied marmots take up residence in rock piles that provide protection from predators. They feed in nearby meadows. Mountain bighorn sheep are occasionally seen. Bird life is similar to that found in other North American mountains, including the White-tailed Ptarmigan (which was introduced).

Formerly widespread, brown bears were exterminated from the Sierra. Mountain lions and black bears can occasionally be seen, but the most common carnivores are small, such as short-tailed weasel and dusky shrew.

***Great Basin.*** Alpine environments are limited to the tallest mountain ranges in the Great Basin, such as the White Mountains, Ruby, and Snake Ranges. Summit areas are small, and the number of plant communities is low. Because the mountain tops are geographically isolated both from one another and from larger ranges, alpine communities have evolved separately as lowland species adapted to changing climate conditions. Many plants are endemic. The western region, in the rain-shadow of the Sierra Nevada, is dry. The east has heavier winter snow and more summer thunderstorms. With the exception of a few late-lying snowbeds or perennial streams, most habitats are dry. Great Basin ranges have not been well studied, but the following generalizations can be made. Mountains east of Elko, Nevada, are more closely related botanically to the Rockies. Alpine floras on the White Mountains and other western ranges are distinct, even though they share several plants with the Sierra Nevada. Treeline is frequently hard to define because of aridity, and the lower limit of the alpine zone is sometimes defined by the absence of woody big sagebrush rather than trees. Wetter areas have a krummholz of white-bark pine, Engelmann spruce, limber pine, or subalpine fir, while rocky or dry open areas have a sparse cover of various grasses. Snow accumulation areas support forb and graminoid plants like mountain hare sedge.

The Ruby Mountains have more precipitation, greater diversity, and greater similarity to the Rockies. Meadow sites are dominated by arctic willow, elk sedge, marsh marigold, and American bistort. Cushions of alpine avens and moss campion account for most of the cover in dry fellfield or rocky ridges. Moist soil around ponds supports Rocky Mountain sedge.

The White Mountains, named for the color of the rock, not snow, are cold and dry with sparse alpine vegetation. Patterned ground, a relic of the Pleistocene, is dramatically apparent with little vegetation to hide it. Although the highest ranges in the Great Basin show effects of alpine glaciation, the White Mountains were too dry for snow to accumulate. Without glaciation, the alpine landscape remains rolling uplands. Of the 200 species in the alpine flora, 62% are also found in the Sierra Nevada, but only 28% also occur in the Rockies. Plant distribution is closely aligned with rock type. Except for stands of bristlecone pine, dolomite (a type of limestone) substrate is dry and barren with little plant cover except mats of Coville's phlox. Granite fellfields have more cover and are dominated by Mono clover. Bottlebrush squirreltail, of cold-desert origin, grows on both substrates, but is more abundant on granite. Three species—phoenicaulis, Mason's Jacob's ladder, and rambling fleabane—are restricted to elevations above 13,000 ft (4,000 m). Many plants are hairy in response to harsh winter and summer conditions. Because the White Mountains receive so little rain and snowfall, no snowbed or marsh communities are found.

Animal life in alpine tundra of Great Basin mountains is limited because of smaller habitat areas. Pikas, deer mice, and yellow-bellied marmots are common. Grasslands and herb-fields also support northern pocket gopher, an indication of affinity with the Rocky Mountains. Isolated areas have rare groups of mountain bighorn sheep. The short-tailed weasel is the common predator of rodents. The variety in bird species is also less than both the Sierra-Cascades and the Rockies, but

· · · · · · · · · · · · · · · · · · · · · · · · · · · · · · · · · · · · · · · · · · · · · · · · · · · · · · · · · · · · · · · · · · · · · · · · · · · ·

### Bristlecone Pine

As its scientific name *Pinus longaeva* implies, bristlecone pine is long-lived (see Plate VIII). The oldest living tree, with 4,844 rings, was cut down on Wheeler Peak in the Snake Range in 1964. Pine Alpha in the White Mountains is 4,300 years old. Interpretive sites exist on the White Mountains and in Great Basin National Park on Wheeler Peak, but bristlecones grow on many high mountains in the Great Basin, usually at elevations of 9,000–11,500 ft (2,750–3,500 m) where summer temperatures barely reach 50° F (10° C), placing it at the edge of the alpine zone. Unlike alpine plants that are of low stature, bristlecones stand exposed to harsh weather conditions—bitter cold, strong winds, blowing ice and snow, and intense radiation. Bristlecones can be large trees or krummholz. Young trees are upright and straight, but old trees subjected to centuries of environmental abuse are broken and gnarled, though still massive. Limestone or dolomite is the preferred substrate. Needles, in bundles of five, can remain on the tree actively performing photosynthesis for 40 years; most other pines replace needles after only a few years. Because of its age and sensitivity to drought conditions that alter the width of growth rings, bristlecones are used in dendrochronology, the science of dating by means of comparing tree rings. Speculation and theories abound as to why bristlecones are long-lived, but no consensus has been reached. Even old plants, 3,000 or 4,000 years old, continue to flower and set seed. Clark's Nutcrackers feed on bristlecone seeds when other food is unavailable and may be responsible for helping to perpetuate the species by caching seeds deep enough that they are protected from the weather, allowing them to germinate and produce seedlings.

· · · · · · · · · · · · · · · · · · · · · · · · · · · · · · · · · · · · · · · · · · · · · · · · · · · · · · · · · · · · · · · · · · · · · · · · · · · ·

Rock Wren, Horned Lark, Clark's Nutcracker, Golden Eagle, and Red-tailed Hawk are seen.

***Rocky Mountains.*** The most studied alpine regions in North America are in the Rocky Mountains, particularly Niwot Ridge and Trail Ridge in the Front Range of Colorado. Because the cordillera was an old and persistent migration route, the Rockies have a rich alpine flora that is similar all the way south to their terminus in northern New Mexico. Alpine sorrel, an arctic species, is one of the most common plants. Typical in alpine environments, a mosaic of communities develops in response to a variety of microhabitats influenced by wind, water, snow, and rocks.

The Southern Rocky Mountains include many ranges in Colorado and New Mexico with a variety of landscape and climate. Even though the mountains have been carved by alpine glaciers, not all of the preexisting uplands were dissected. Large areas such as Trail Ridge in Rocky Mountain National Park remain as undulating expanses above treeline. The San Juan Mountains in southwestern Colorado, directly in the path of west winds and cyclonic storms, get both heavy snows in winter and thunderstorms in summer. The Sangre de Cristo range in New Mexico is isolated from the main cordillera but has similar plant communities. Arctic plants are usually limited to wet marshes in the Southern Rockies and are smaller than the same plants found in the Arctic. Mountain avens and kobresia, common in the northern ranges and in the

## Glacial Water

Boulder, Colorado, located on the semi-arid Great Plains at the base of the Colorado Rocky Mountains, is unique among American cities in that it gets almost half of its water supply from a glacier. Early in the twentieth century, the city purchased Silver Lake, where meltwater accumulated before flowing into Boulder Creek and through town. In the 1920s, the Burlington Railroad promoted visits to the Arapaho Glacier as a tourist excursion, drawing hundreds of people from the Chicago area to play on the snow and ice. Alarmed at the potential despoilment of their water supply, the City of Boulder purchased 3,869 acres of land from Roosevelt National Forest, including the Arapaho Glacier and its entire watershed, for $1.25 per acre. To maintain purity, the watershed is now closed to the public, and the surrounding alpine environment remains unspoiled. Permits may be obtained for scientific studies, however.

## INSTAAR

The Institute of Arctic and Alpine Research (INSTAAR) is an interdisciplinary institute in the Graduate School at the University of Colorado. Students in many fields, including anthropology, ecology, geography, geology, atmospheric, and oceanic sciences, study and conduct research on how the Earth's surface is affected by natural and human-related processes. Programs focus on ecosystems, geophysics, and past global change. Faculty have special expertise in polar and alpine research on such topics as arctic and antarctic hydrology, atmospheric dynamics, alpine ecology and climatology, invasive species, arctic climate change and variability, geochronology, and paleo-ecology. Understanding environmental processes is a prerequisite to dealing with world problems, such as maintenance of water quality and consequences of long-term environmental alterations. Field-based courses and research opportunities are available at the Mountain Research Station, 20 miles west of Boulder, and the Institute maintains a meteorological station on Niwot Ridge at 12,280 ft (3,743 m).

Arctic, are almost absent. The Central Rocky Mountains, extending from Colorado to Montana, also have a variety of landforms and rock types, and communities vary accordingly. Calcareous soils in Glacier National Park, for example, support communities similar to dry sites elsewhere because limestone does not retain water.

Principal treeline species in the Rockies in the United States are Engelmann spruce and subalpine fir. In Montana, limber pine grows on drier sites, while whitebark pine occupies more moist sites. In the Canadian ranges, alpine larch and western hemlock grow with an understory of heath shrubs in wetter climates similar to the Cascades. Treeline in Alaska and the Yukon resembles the transition from boreal forest to arctic tundra, where white spruce, black spruce, and tamarack dominate.

Talus and scree slopes are stabilized by subsurface rhizome networks of several forbs, including milk vetch, drabas, and ragworts. Plants in boulder fields include many saxifrages and alpine sorrel. Because they depend on sparse snowmelt in largely snowfree areas, plants in fellfields bloom early. Rosettes of alpine primrose, cushions or mats of alpine phlox, alpine candytuft, moss campion, dwarf clover, and many others dot the ground with colorful flowers (see Plate IX). Although sedge meadows dominated by kobresia tussocks are limited to areas blown free of snow during winter, exposing the plants to weather extremes, they contain more species than any other alpine community. Other plants include one-flowered harebell, arctic gentian, and several more tiny plants. About one-fifth of the plants are lichens, the most abundant being Iceland lichen, which is widespread in windy, snowfree areas in the Arctic and throughout the alpine in North America. One of the most common plants in snowbeds is the snow buttercup, blooming at the edge of the snow as it melts. Prostrate, creeping arctic willow also grows near snowbed margins. Tussocks of tufted hairgrass, found in alpine regions in many parts of the world, form a meadow-like community in the snowbed. Bistort and creeping sibbaldia are also common. The center of the snowbed remains saturated all summer and supports clumps of Drummond's rush. All snowbed plants grow in precise locations according to the time of melt, which determines the length of the growing season.

Plants that flourish in soils disturbed by pocket gophers tunneling beneath the snow cover include sky pilot, alpine avens, and alpine sage. Many of these pioneering plants are large and showy, attracting bees and flies for pollination. Alpine marsh environments with saturated soils and standing water are green spots in frequently dry tundra. Marshes often occur where water flows out of a tarn. Although Rocky Mountain sedge dominates, other sedges are also found. Succulent sedums such as rose crown are common. Succulence is an adaptation to conserve moisture and is beneficial because plants have difficulty obtaining water from cold acidic soils. Marsh marigold and cottongrass are common in wet areas. Several rushes grow in the gravelly streams, and several types of willow grow in thickets along stream channels.

The Northern Rocky Mountains in Alberta are primarily sedimentary or metamorphosed sedimentary rocks. Vegetation varies widely according to geographic location, frost action, microhabitat, and substrate. In contrast to the southern and central Rockies, the dominant growthforms are dwarf shrubs, some evergreen such as bell heather, and others deciduous species such as willows. Moist areas support grasses and sedges, while rosettes and cushion plants occur on drier ridges. Rocky and heath tundra, the driest sites, are dominated by mats of mountain avens, arctic willow and arctic bell heather, with rosettes of alpine locoweed. Several mat-forming plants more typical of arctic climates include black crowberry and mountain heathers. Elk sedge and mountain hare sedge tussocks form meadows. Meadow and snowbed plants have the shortest growing season but need both winter protection and summer water.

Arctic and alpine zones in Alaska and the Yukon merge, and the alpine environment and flora resembles that of the Arctic. The east-west trending Brooks Range in Alaska at 69° N is a climatic barrier preventing northern airmasses from penetrating farther south, separating arctic tundra from boreal forest in interior Alaska. A spruce treeline exists on the south slope, but there is no forest on the north. A mosaic of vegetation exists according to slope steepness, aspect, soil depth, and texture. The highest exposed elevations have a reindeer lichen turf with moss. Rocky summit areas, boulder fields and fellfields, are limited to primarily northern woodrush and dotted saxifrage. The most common community on drier ridges is composed of sparse cushions and mats, 90% of which are mountain avens. Deeper soils downslope have downy birch, willow, and heath shrubs. Gentle slopes with fine soils, impeded drainage, and thick organic horizons have sedge meadows or cottongrass tussocks similar to that found in Low Arctic tundra. Dwarf shrubs of marsh Labrador tea and bog bilberry also occur, and mosses account for 35% of the cover.

Several associations occupy snow accumulation sites. The center of the snowbed with the shortest growing season has forbs of Bigelow's sedge and bistort. Areas with a longer growing season have arctic bell heather surrounded by a lichen-heath community. Where snow melts first, a mountain avens and willow fellfield develops.

Mule deer and pronghorn may be occasional visitors, grazing wet meadows, willow, and krummholz. Bison and elk formerly occurred at high elevations in several mountain areas, but most have been exterminated. However, elk have been reintroduced in some places. Bull elk occasionally remain on the tundra all winter, grazing in open meadows.

Mountain goats are native only to the mountains in southeastern Alaska, Canada, and western Montana (see Figure 3.10). Mountain sheep are plentiful in remote areas of the Rockies. The all-white Dall's sheep, distinguished by its thin horns, ranges from Alaska to northern British Columbia, while mountain bighorn sheep is found in the Rockies in the United States and occasionally in the Great Basin. Most bighorn sheep in Colorado live on the eastern slope where snowfree rocky crags provide winter forage. Rams gather a harem of ewes. Lambs, born in May and July, are capable of jumping and climbing cliffs when only three days old.

Rodents are widespread. Dusky shrew, deer mouse, heather vole, and meadow vole are found in a variety of habitats. The most common small mammal in moist

**Figure 3.10** Mountain goats are native to the northern Rocky Mountains. *(Photo by author.)*

meadows is the meadow vole, in willow thickets it is the dusky shrew and water vole, while krummholz is home to yellow-bellied marmot and heather vole. Pikas live in rock piles. The collared pika occurs in the Yukon and Alaska, and the American pika occupies similar habitats in Alberta, British Columbia, and the United States. The two most common marmots both live in the Rocky Mountains. The hoary marmot, native from Alaska to northern Idaho, is replaced by the yellow-bellied marmot in most of the Rockies in the United States. The Alaska marmot is limited to northern Alaska.

Brown bears are still common in the remote Northern Rockies and as far south as Glacier and Yellowstone National Parks in Montana and Wyoming. Short-tailed weasels hunt rodents, and dusky shrews eat small invertebrates.

Bird life, including White-tailed Ptarmigan, Water Pipet, Rosy Finch, Horned Lark, and Golden Eagle, is typical of most mountain areas in western North America.

Although most amphibians and reptiles are restricted by cold temperatures, some boreal forest toads reach high elevations in the southwestern Rocky Mountains, as do a few salamanders in Colorado. Along with common pollinating insects, such as bumblebees and flies, four species of ants inhabit the Colorado alpine.

***Mount Washington and other high peaks of eastern North America.*** The Appalachian Mountains in eastern North America are older than western ranges and have rounded topography and lower summits. Plants are more closely related to the Arctic than to the Rocky Mountains. Arctic flora that had been pushed south in advance of the Pleistocene ice sheets took refuge upslope as the climate warmed and glaciers retreated. Alpine environments now exist only on isolated summits that were covered by continental ice during the Pleistocene, including Mount Katahdin (Maine), the White Mountains (New Hampshire), Green Mountains (Vermont), Adirondacks (New York), and the higher peaks in Canada, such as the Notre Dame Mountains on the Gaspe Peninsula and the highlands of southern Labrador and eastern Quebec. Although elevations are not high compared with mountains in western North America, cold, snowy, or foggy conditions depress treeline to lower elevations. As in other alpine areas, the mean temperature of the warmest month is below 50° F (10° C) and the summer growing season lasts four to five months. Although dry microclimates exist, the overall climate is moist, with snow or rain all year and many foggy days. Mount Washington, in the White Mountains, averages 90 in (2,280 mm) of precipitation a year and frequently experiences foggy or icy conditions.

Patterned ground that developed during the Pleistocene and is no longer active can be seen on soils formed from crystalline rocks of schist and gneiss. Because of extreme variation in weather conditions, krummholz or the treeline zone has a wide elevational range, 4,800–5,300 ft (1,450–1,600 m), and stunted trees can often be found in the shelter of large boulders. Several plant communities, typified on Mount Washington, are present, reflecting the interrelated environmental factors of fog, snow accumulation, and wind. Because of persistent foggy conditions, treeline species are distinct from those of both the western mountains and the Arctic. Balsam fir dominates, but black spruce, found only in bogs in the Arctic, and groves of paper birch are prominent. Flagged trees indicate strong prevailing wind patterns. Even though windy, summits with heavy fog and deep snows are moist all summer and support sedge meadows dominated by Bigelow's sedge, haircap moss, and Iceland lichen. Even porous sandy soils remain moist because of near constant foggy conditions. Melting free by early July, snowbank communities under heavy fog are the most species rich with many heath and willow shrubs, along with nodding hairgrass.

Several different shrubby heath communities exist in areas with less moisture, either due to fewer foggy days or less snow to provide meltwater; indeed, they are the dominant vegetation type in the alpine zone of the White Mountains (see Figure 3.11). Bog bilberry is the most common shrub in eastern alpine regions, joined by mountain cranberry and sweet blueberry.

Other low plants such as Labrador tea shrubs, Bigelow's sedge, highland rush, and Greenland sandwort and Lapland diapensia cushions vary with locale. Club-mosses grow protected beneath the shrubs. Windswept areas without snow have a sparse cover of diapensia cushions, with highland rush and some low-growing bog bilberry. Sunny streamsides protected from wind have marshy vegetation, shrubs of bearberry willow, mountain alder, and three-toothed cinquefoil, along with

**Figure 3.11** Low heath shrubs protected among rocks are typical of Mount Washington and other eastern North America mountains. *(Photo by author.)*

sedges and viviparous bistort. Bogs in similar locations are limited in extent and dominated by Bigelow's sedge. Because foggy conditions filter solar radiation and moisture is available all year, these mountain areas have few hairy or succulent plants. The dominant heath and lichen vegetation on Mount Washington closely resembles that of the Arctic. About 70% of the vascular plants on Mount Washington are of arctic origin. Fruticose lichens, also typical of the Arctic, are more common in the cool, wet environment than crustose forms more prominent in drier areas.

Animal life on high mountains in eastern North America is limited. Typical species include deer mice, meadow voles, and groundhogs, solitary animals that are also called woodchucks. Related to marmots, groundhogs grow to be 20 in (50 cm) long and weigh less than 6 lb (2.7 kg). Like marmots, they prefer rocky habitats adjacent to meadows. Short-tailed weasels are common, as is the masked shrew. Frequently seen birds include Slate-colored Juncos and White-throated Sparrows. Red-tailed Hawk is the most common raptor.

### Europe

The east-west orientation of most Eurasian mountain ranges allowed few migration corridors to develop between mountain tops and the Arctic (see Figure 3.12). As a consequence, Eurasian alpine communities contain fewer arctic plants than

···········································································

**Extreme Weather**

Although the summit of Mount Washington is only 6,288 ft (1,917 m), it is famous for weather that can change from warm and sunny to freezing fog and gale-force wind within minutes. More than 100 people have died on its summit or slopes, many during the summer months. The extremes result from a unique set of circumstances including a north-south orientation and a position beneath the jet stream where three storm tracks converge. Wind speed in summer averages 25 mph (40 km/hr), increasing to 45 mph (72 km/hr) in winter. However, 100 mph (160 km/hr) winds can be expected about every three days in winter, and the highest wind ever recorded on Earth's surface, 231 mph (372 km/hr) occurred on Mount Washington in April 1934. Buildings and large instruments are chained down to prevent their loss in high winds. Even without wind, tempera-tures are consistently cold. Mean annual temperature is only 26° F (−3° C), with a winter extreme of −47° F (−44° C). Wind chill may lower the perceived temperature to −120° F (−84° C). The highest temperature ever reached in summer is only 72° F (22° C). Snowfall is high, averaging 255 in (650 cm) per year, even though it is often redistributed by strong winds. The maximum ever recorded for one season was 566 in (1,438 cm). The summit is foggy more than 300 days of the year. If the cold air is clean and lacks freezing nuclei (particles for ice to cling to), it will be super-cooled, meaning that water droplets are unable to change state into solid even though air tem-perature is well below freezing. When supercooled fog comes into contact with buildings or instruments, the water droplets instantly freeze, coating the objects with a thick coat of rime ice (see Plate X). Weather records were taken 1870–1892, but the summit was unoccupied until 1932. An independently operated Mount Washington Observatory now houses scientists and volunteers who take weather measurements and do research year-round on such diverse topics as ice physics and cosmic rays.

···········································································

similar habitats in North America. Most plants and animals, however, are close rel-atives of North American forms. Even with few arctic species, dominant growth-forms are the same as those in arctic tundra.

*European Alps.* The European Alps extend only a short distance north to south, 44°–48° N, but cover a 750 mi (1,200 km) arc from France in the west to Austria in the east. The widest point is only 150 mi (240 km). Alpine conditions exist above roughly 6,500 ft (2,000 m). The highest peaks are Mont Blanc in the west (15,771 ft, 4,807 m), Piz Bernina in the middle (13,284 ft, 4,049 m), and Grossglockner in the east (12,457 ft, 3,797 m); all have alpine glaciers. Like the Himalayas, Andes, and Rockies, the Alps are young mountains, dating from the Tertiary. High plateaus, which form step-like levels at 5,900 ft (1,800 m), 8,200 ft (2,500 m), and 9,850 ft (3,000 m), are old erosion surfaces produced during geologic pauses in mountain uplift. During the Pleistocene, when ice almost completely covered the Alps, valley glaciers sharpened the landscape and deposited extensive stony moraines in val-leys. Calcareous bedrock and soils are characteristic of the outer chains, while

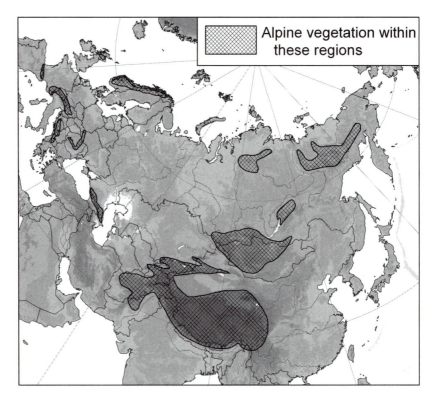

**Figure 3.12** Alpine tundra in Eurasia. *(Map by Bernd Kuennecke.)*

granitic materials dominate in the inner mountains. All soils are young, reflecting characteristics of the substrate, and closely correlate with vegetation. The Alps have been heavily influenced by human activities for thousands of years, so it is difficult to say what makes up the natural vegetation.

The European Alps, as well as mountains such as the Pyrenees, Carpathians, Caucasus, and northern Balkan Peninsula in Europe, rise between two biomes: the Temperate Broadleaf Deciduous Forest to the north and the semi-arid Mediterranean Biome to the south. No other mountains in the world occupy similar positions. Half of the mountain flora is derived from the southern and central Europe lowlands, while the other half is dispersed from the north and east. No endemic plant families and only one endemic genus (*Berardia*) are found in the Alps. However, bellflower, draba, lousewort, primrose, buttercup, saxifrage, and violet genera contain many endemic species. Some endemics are distributed throughout the Alps; others are concentrated in refugia. About one-half of the plants are restricted in distribution, while the other half migrated from the Arctic or from central Asian highlands since the Pleistocene.

Mountain climate varies from the outer regions to the inner valleys and also from east to west, but generally has enough rain in warm periods for plant growth and is cold enough in the winter for plant dormancy. Both north and south slopes are wet—rain in summer and snow in winter—with annual precipitation up to 100 in (2,500 mm). Parts of the central mountains have continental climate characteristics with temperature extremes. Plant communities occur at different elevation levels according to location, at the fringes or interior of the mountains.

Treeline species in the outer regions are Norway spruce or prostrate pine. The inner and central Alps have mixed forests of European larch and stone pine or forests of mountain pine. Natural vegetation on steep slopes usually has not been disturbed by human activities, but due to agricultural clearing for hundreds of years, it is impossible to tell where natural treeline is in most of the Alps. Forests on steep slopes are cut through by avalanches and rock slides. Huge treeless rock outcrops, especially in the limestone areas such as the Dolomites of northern Italy also complicate the issue of treeline (see Plate XI).

The alpine area has a lower zone with dwarf-shrub heaths, an upper zone with dense, steppe-like grasslands or meadows, and a subnival zone. Some of the dwarf-shrub heath in the lower zone may be undergrowth of forests that have been cut, but there is no way to tell the difference between those remnants and alpine vegetation. Both are thickets with dense moss or lichen layers beneath. Dominant shrubs may be rhododendrons, blueberries, or creeping azalea, depending on microhabitat. Snowbed communities do not occur in the lower zone because snow melts too quickly, but wetter areas support matgrass in their centers with dwarf shrubs such as rhododendrons at the edges. Frost action prevents a closed vegetation cover on exposed ridges.

The beginning of the upper zone is where alpine grasslands form dense carpets and replace dwarf shrubs. The grasslands, however, are primarily sedges because of dry habitat. Different species of carex grow according to the acidity of substrate. Associated with them are rosettes with rhizomes, woody creeping mats, and tiny dwarf shrubs. Fruticose lichens grow between sedge plants. On windswept ridges without a protective snow cover, kobresia sedge replaces carex. Ridges may experience temperatures down to −58° F (−50° C), which creeping azalea, purple saxifrage, and moss campion can withstand. In contrast, microclimate conditions in summer can be close to 100° F (37.8° C). Houseleek grows where the surface is hot and dry, but the succulent plant suffers no drought stress. On steep, sunny slopes, fescue grasses replace sedges.

Under prostrate green alder, which grows mostly in avalanche paths, the soil is rich in nitrogen, so forbs grow larger than normal. On disturbed areas such as scree, which are susceptible to movement, major growthforms are creeping mats and gravel stabilizers because loose rock surfaces provide only holes or crevices for rooting. Spherical cushions of rock jasmine grow here, needing little room for root growth. Lichens and mosses often occur in large numbers. Edelweiss can grow on both scree slopes and in meadows. Norwegian hairycap moss occurs in the wettest areas of snowbeds with the shortest snowfree period, while creeping willows grow at the edges. Snowbed plants need protection from frost in winter, and many plants cannot survive if the temperature drops too low. Sensitive plants have affinities to the warm temperate or Mediterranean zones. Frost-tolerant plants usually have an arctic or northern high mountains origin.

The subnival zone, a transition to the region of permanent snow, is characterized by patches of grassland and scattered cushion plants. Within this zone, plants grow in sheltered, sunny niches. In Ötztal, Austria, the permanent snow line is 10,170 ft (3,100 m), but several plants grow as high as 11,480 ft (3,500 m). Formerly thought to be the highest vascular plant in the Alps, glacier buttercup at 12,073 ft (3,680 m) was later discovered to be topped by two-flower saxifrage growing at 14,600 ft (4,450 m). Mosses, lichens, bacteria, and algae grow even higher.

Red deer and chamois are common herbivores; they may overpopulate some areas and damage subalpine forest. In contrast, mountain ibex is rare and only found in the Alps, although related species occur in the Pyrenees and Caucasus Mountains. Ibex, formerly widespread, had been reduced to a small herd in the Gran Paradiso area in northwestern Italy, but now has been successfully reintroduced to other parts of the Alps (see Figure 3.13). Marmots are the most conspicuous rodent. Both snow vole and alpine shrew are endemic to the Alps. The snow vole's nearest relative occurs in Eastern Siberia. The presence of endemics illustrates the relative isolation of European ranges, while affinities with eastern Asian species reflect east-west migration routes. Most of the larger carnivores, such as

**Figure 3.13** Ibex, seen here near Gran Paradiso in the Italian Alps, are native to many mountains of Europe and Asia. *(Courtesy of Shutterstock. Copyright: ArturKo.)*

brown bear, lynx, Bearded Vulture, and Golden Eagle, have been hunted to near or actual extinction. Golden Eagles are now protected. Small predators include the insectivorous alpine shrew and the wildcat.

The European Alps had already been extensively used prior to Roman conquest, with many permanent settlements built close to treeline. Summer grazing took place on alpine meadows, and hay was harvested on slopes too steep for grazing. Conversion of forest to pasture on steep slopes increased the frequency and area of avalanches. Approximately one-half of current avalanches can be attributed to human activities. European countries now use alpine areas in different ways. Some continue traditional agriculture, while others have abandoned the land. Tourism and recreation, especially the creation of ski runs, which destroys vegetation and soil, have had profound effects.

## Central Asia

Except for the Himalayas and adjacent ranges, the high mountain systems of Asia, are generally located in the southern part of the former United Soviet Socialist Republic. The Carpathians, Altai, Sayanskiy Range, and Tarbagatay Ranges rise

........................................................................

### The Pyrenees

Separating the Iberian Peninsula from the rest of Europe, the Pyrenees Mountains are a barrier to cyclonic storms coming from the northwest. The French side of the mountains receives the brunt of the storms and is cool and snowy, while the Spanish side in the rainshadow is warm and dry. The difference in climate is reflected in the vegetation. White fir and European beech forests cloak the northern slopes, contrasted with shrubs and grasses on southern slopes. Both sides, however, have deep glaciated valleys and active glaciers. Pico de Aneto, 11,168 ft (3,404 m), is the highest point. Much of the flora and fauna are similar to the Alps, but many endemic species, such as Pyrenean saxifrage, iris, and blue thistle, occur. The Pyrenean chamois, or isard, is slightly smaller than chamois in the Alps. Marmots have been successfully reintroduced, and genetically similar Slovenian brown bears were introduced to supplement the dwindling population of Pyrenean brown bears.

........................................................................

........................................................................

### Cairngorm Mountains

The highest peaks in the Grampian Mountains, including the Cairngorms, in central Scotland are Ben Nevis at a mere 4,406 ft (1,343 m) and Ben Macdui at 4,295 ft (1,309 m). Despite the relatively low elevation, however, remnants of tundra vegetation, relics from the Pleistocene, grow on many slopes and summits in the range. Great Britain was almost completely covered by glaciers during the Pleistocene, and ancestors of mountain plants survived conditions at the edges of the ice. Due to a warm ocean current offshore, today's climate is moderate for the mid-latitudes and is similar to maritime tundra on arctic coasts. Many of the alpine plants are also found in the Arctic. Birds include several species, such as ptarmigan, that are also found on arctic tundra.

........................................................................

## Carpathian Mountains

The maritime influence on the Carpathians, from Slovakia east through southwestern Ukraine and south to Romania, is illustrated by its high annual precipitation, 55 in (1,400 mm), and mild winter and summer temperatures. Because the mountains are surrounded by forest rather than desert, about one-half of the alpine plants are derived from boreal or deciduous forest elements. The highest peak, Mount Goverla, is only 6,762 ft (2,061 m), and variation in treeline elevation is due to human activities. Natural vegetation is dominated by pines and rhododendron shrubs, with tall herbs that thrive with plenty of water. Heath and mosses are also typical. Natural shrubby communities are replaced by secondary growth meadows where human influence has been high. When overgrazed, other grasses are replaced by matgrass, the final stage in human-induced change.

above temperate forest, while the Tien Shan is surrounded by deserts. The Tien Shan and Carpathians have been most studied, the former being more continental while the latter has more maritime influence.

*Tien Shan Mountains.* The Tien Shan, at 42° N, are northeast-southwest trending ranges that form the border between Kyrgyzstan and the Tarim Basin in China. Average elevation of the alpine zone is 13,450 ft (4,100 m), and some peaks rise more than 16,400 ft (5,000 m). Part of the range is a plateau at 12,500 ft (3,800 m). Because all the high mountains were glaciated and piedmont glaciers covered the plateau, moraines and outwash are common. Continuous permafrost exists above 10,800 ft (3,300 m). About 70% of the total species in the Tien Shan alpine zone originated in the Pamir-Tien Shan region or other mountains of southern and central Asia. Arctic flowering plants are only 10–20% of the total and are more common in moist communities. Few plants have affinities to mountain floras of southern Europe.

In this harsh environment, due not only to high elevation, but also relatively low latitude and surrounding aridity, solar radiation is intense. Temperatures and length of growing season are comparable to parts of arctic tundra. Air and soil temperatures drop below freezing even on most summer nights. The lower limit of the alpine zone is 9,500 ft (2,900 m), where mean July temperatures are 50° F (10° C) and mean January temperatures are −4° F (−20° C). Growing season ranges from six months at the lower limit to only one-and-a-half months at the upper limit (13,100 ft, 4,000 m). Winds are strong in summer due to surface heating. Nights, however, are calm due to radiational cooling and subsiding air. Winters are also less windy because of dominant high pressure caused by cold air. The mountains are dry, with as little as 6 in (150 mm) of precipitation per year. Even in the wetter summer season, precipitation may fall as hail or snow. Winter snow cover, which lasts two to six months, is thin and unevenly distributed. Below 11,500 ft (3,500 m), communities are xerophytic, but higher elevations are more humid. In the subnival zone, mean monthly temperatures rarely rise above freezing, and the zone of permanent snow and ice begins at 14,100 ft (4,300 m).

Because the mountains rise from desert, there is no subalpine forest. Only a sparse treeline of central Asian spruce exists on the wetter northern slope. Several alpine communities, with overlapping limits, are based on elevation, temperature, and precipitation. Soil moisture is most significant. The two lowest alpine zones, best developed in the eastern region, are desert and semidesert with about 8 in (200 mm) annual precipitation. Both have salty soils low in humus and basically vary in terms of the amount of grass cover. In the desert zone, at about 10,000 ft (3,000 m), the combination of aridity and poor soils supports only a thin cover of sagebrush and salt-tolerant shrubs with a few sea lavender herbs and low shrubby cushion plants. Slightly wetter semideserts are still dominated by sagebrush, but also have needle-grass and a few other flowering plants.

Higher on the slopes are dry steppes and cold steppes. Soils tend to be saline. Dry steppes, with 9 in (230 mm) of precipitation, are dominated by tussocks of fes-cue, spike fescue, and false needlegrass. Few nongrasses occur, but lichens are common, especially crustose and fruticose. Cold steppes, both slightly wetter and cooler, have a distinctly different community dominated by aster-like snow lotus, with locoweed at higher elevations or fescue, reedgrass, and false needlegrass at lower elevations. Other flowering plants are sparsely represented.

Higher elevations are more moist due to more precipitation and cooler temper-atures. Steppe meadows receive 11 in (280 mm) and have a four-month growing season. Snowmelt provides sufficient moisture content in spring, but soils become dry in summer. Communities are dominated by kobresia sedge or Tien Shan fes-cue. Because of summer drought, few forbs appear, although asters, gentians, dan-delions, and some others can be found. Slightly higher and wetter, moist meadows with deep soils (10 in, 25 cm) are widespread, particularly on shady slopes and river terraces (see Figure 3.14). Most typical is a kobresia and carex sedge

**Figure 3.14** Lush meadows in the Tien Shan Mountains, Kyrgyzstan, are surrounded by high, snowy peaks. *(Photo by author.)*

community that includes 75 flowering plant species in several associations. Shag-spine pea-shrub and alpine bluegrass dominate. Many fruticose and foliose lichens and several mosses grow on the soil surface. Where frost polygons limit vegetation cover, plants are androsace cushions, buttercups, snow lotus, and crustose lichens. These extensive wet sites are the basis for a rather complex ecosystem of plants, herbivores, and carnivores.

The highest alpine zone below permanent ice and snow, occurring at elevations of 13,000 ft (4,000 m), is an herb tundra and cushion plant community similar to that found in the Arctic. Plants do not form a continuous cover on the shallow, stony soil, but grow in patterns similar to frost boil tundra. Cushion plants of *Dry-adanthe tetranda* dominate but cover only 10–30% of the ground. Most other flowering plants and mosses grow within the cushions.

Bog communities can be found at any elevation in the alpine zone where groundwater reaches the surface. Soils are peaty glei with organic content up to 60%. All bogs are characterized by sedge communities dominated by carex species. Forbs include lousewort, viviparous bistort, and buttercup. Mosses grow on the soil surface.

Although many plant species are restricted to particular communities, animals in the Tien Shan tend to be more widespread, occupying several elevation zones and habitats. Gray marmots and Tolei hares are common in all communities except the highest moist meadows and herb tundra. Marmots are abundant wherever it is rugged and rocky. Gray hamsters occur only in lower elevations, in desert to dry steppe. Large-eared pika, southern mole vole, and silvery mountain vole, limited to higher elevations, live in habitats from dry steppe to moist meadows. Voles and marmots are the most characteristic mammals of the moist meadows. Ungulates that graze the meadows are Siberian ibex and the argali or Kyzylkum sheep. Carnivores include weasels, fox, and an occasional brown bear or wolf, especially at the higher elevations. Birds, including Horned Lark, Lesser Sand Plover, Isabelline Wheatear, and Snow Finch, are abundant in dry communities. White-winged Redstart and Alpine Chough are also common. In bogs, voles are the only permanent inhabitants and Water Pipet is the only nesting bird, although other animals and birds feed in bogs.

### Speciation in the Himalayas

Because of their subtropical latitude, the Himalayas have little floristic relationship with circumpolar tundra. No woody perennials are common to both regions, but the Himalayas have 20 herbs of circumpolar distribution, including viviparous bistort, a carex sedge, fireweed, kobresia, Iceland purslane, and alpine sorrel. The cryptogam flora is less known. Fruticose species of Iceland lichen, reindeer lichen, and white-worm lichen dominate but do not form extensive carpets. The range supports a distinctive diversity of alpine plants. Only six species of saxifrage occur in the Arctic, but 100 grow in the Himalayas, Similarly, one primrose grows in the Arctic compared with 90 species in the Himalayas. The Arctic has only two species of rhododendron and six of saussurea, while the Himalayas have 40 rhododendrons and 60 saussureas.

*Himalaya Mountains.* At approximately 30° N, the isolated Himalayan mountain range contains the southernmost and highest alpine ecosystem of the world. In a southward curving arc, the range covers a large east-west extent from the Indus Gorge in northern Pakistan to the Tsang Po Gorge in northeastern India. Most of the Himalayas are composed of acidic rocks, such as granite and gneiss, and soils are young regosols. At elevations below 13,000 ft (4,000 m), the landscape is characterized by U-shaped glacial valleys with moraines on higher slopes. Valley floors are filled with muddy outwash material and alluvial fans that developed on steep side walls after valley glaciers retreated. Solifluction occurs above 13,000 ft (4,000 m). Above 14,750 ft (4,500 m), debris-covered glaciers fill valley floors between steep slopes (aretes) that rise above the ice. Because of steep slopes and frost action, scree slopes are plentiful, providing habitat for pioneer plants. The lower alpine zone is characterized by closed vegetation on all slopes. In the higher alpine zone, sedge mats and cushions exist in the shelter of large rocks or on patterned ground. The highest plants in the world grow in isolated protected sites on the slopes of Mount Everest and surrounding peaks.

The alpine environment varies according to geographic location and aspect— north-facing and south-facing slopes are significantly different. Treeline elevation, as well as species, is highly variable but generally 12,000 ft (3,600 m). On shady slopes in the northwest, treeline species are fir, spruce, and pine. On sunny slopes in the same locality, sea wormwood dwarf shrublands and Grecian juniper woodlands merge with subalpine meadows. In the Central Himalayas, cloud forests of east Himalayan fir, birch, and drooping juniper at the upper treeline are replaced by subalpine thickets of rhododendron on shady slopes and juniper on sunny slopes. In the drier Inner Himalaya, because it is more sheltered from monsoon rain, closed dwarf-shrub thickets and mats are replaced by the open dwarf shrublands and steppes of the Tibetan Plateau.

No long-term weather stations exist, and the climate of the Himalayas is not well known. Vegetation and data from short-term weather measurements indicate that the entire range is humid, but especially so in the southeast. Even in the drier northwestern part, dwarf sage shrubs form a closed vegetation cover with many forbs. From northwest to southeast, summer rains increase while winter and spring rains decrease. The northwest gets winter precipitation from cyclonic storms moving in from the west; only one-quarter of its annual total falls in summer. Plants there grow best under or around winter snowbeds, while others depend on light summer rain. The southeast, dry in the winter due to monsoon winds from the Asian interior, receives most of its precipitation in the summer from southerly monsoon flow. The highest amount of precipitation falls in the cloud forest below the alpine zone. Although it receives lower precipitation, the alpine belt has high humidity in the summer, which promotes growth of foliose lichens and diversity in floristic composition. A pronounced difference in microclimate is evident on the north and south slopes in the eastern Himalayas during the dry winter. South-facing slopes are hot, while north-facing slopes are shady and snowy until the end of May. The difference diminishes in the summer when both slopes are moist due to high humidity.

The lower alpine zone, near 14,000 ft (4,250 m), gets mostly rain, and a long-lasting cloud cover suppresses diurnal differences in temperature and humidity. In the upper alpine, at 15,500 ft (4,750 m), weather is dominated by drizzling mists alternating with sunshine. This causes short-term changes in temperature and relative humidity. Surfaces are always damp and covered by more lichens than flowering plants.

At the lower limit of the alpine zone, more than 200 days have a mean temperature around freezing. The upper alpine zone is more continental, with more extreme temperatures, especially during the wet summer monsoon when the cloud layer is either thin or well below this high zone. Temperatures in the upper alpine regularly fall below freezing during the wet summer monsoon season, causing needle ice. Between the rainy season and the first snowfall, some plants like gentians, snow lotus, and larkspur can withstand temperatures down to 5° F (−15° C) at night. According to estimates from existing data, the mean summer temperature at the upper limit of the alpine belt is 36.5° F (2.5° C). The extent of permafrost is not known, but extensive arctic-size areas of patterned ground can be seen above 16,500 ft (5,000 m). The elevation of the open alpine mats at the upper limit of the alpine zone may coincide with the lower limit of permafrost.

Both wind and duration of snow-cover affect plant communities. Sheltered areas have less temperature variation, while windy ridges are more extreme. High ridges exposed to cold, dry winter winds may be completely blown free of snow. Evergreen woody perennials like dwarf-shrub rhododendrons are restricted to snow-protected slopes.

Because the Himalayas cover such a large area and have a wide range in temperature, precipitation patterns, glaciation, substrate, slope steepness, and slope aspect, there are many plant communities depending on geographic location. Although the growthforms of the basic communities are similar throughout the range, species composition varies.

The two major communities of the lower alpine zone are both dwarf-shrub thickets (see Figure 3.15). Rhododendrons grow as cushion-forming dwarf shrubs on both sunny and shady slopes up to 14,750 ft (4,500 m) where they are protected by winter snow. Tall forbs such as lousewort, monkshood, and kobresia sedge grow throughout the woody evergreen thickets. The humid conditions support abundant lichen growth in the shelter of the rhododendrons. Several junipers form a second set of dwarf shrublands in the lower alpine, some restricted according to geographic area. Single-seed juniper is found throughout the Himalayas, common juniper in the northwest, and Grecian juniper on the driest rocky outcrops. Some form flag trees due to strong upslope winds. Juniper roots stabilize soil on steep slopes better than rhododendrons, even in areas affected by solifluction. Between the juniper patches are mat-forming meadows of Nepal kobresia sedges and flat, silvery-colored cushions of everlasting and edelweiss. Mosses and lichens are rare.

Rhododendron-kobresia mosaics are the major vegetation in the upper alpine zone at 15,000 ft (4,600 m). Dwarf shrubs, common in the lower alpine, are limited

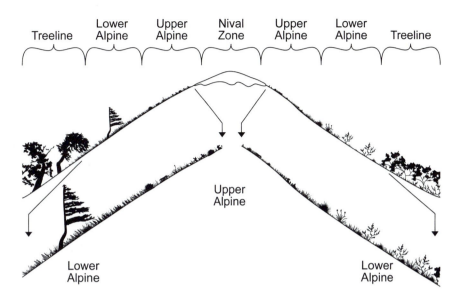

**Figure 3.15** Vegetation varies with elevation as climate and environment changes in the Himalayas. *(Illustration by Jeff Dixon.)*

to protected warmer habitats in the upper alpine. Other plants in the kobresia turf include plants with long taproots, both rosettes and cushions, such as rock prim- rose. Except for their flowering stalks, plants are less than 2 in (5 cm) high, and the leaves of the cushions or rosettes are tight to the ground. On drier and windier sites susceptible to disruption by solifluction, only small patches of flowering plants such as taprooted cushions of snow lotus, sandwort, and sibbaldia grow. Mosses and lichens outnumber flowering plants. Crustose and foliose lichens cover dead kobresia plants, and fruticose lichens grow on woody parts of rhododendrons. Gravels between plants are also covered with lichens. Open scree slopes in the northern part of the range have cushions and rosettes with long roots that anchor plants in the unstable ground. The drier north slope is an alpine short-grass steppe with needlegrass, carex sedges, and flat cushions of rock primrose and locoweed.

Two major plant communities depend on variations in soil. Snowbeds, similar to those found in the European Alps and other mountains in North America and Eurasia with winter precipitation, are more common in the upper alpine zone in the northwestern Himalayas. In this summer-dry region, melting snow provides a needed source of moisture in summer. Bistort, buttercup, and mosses dominate. Vascular plants are rare on unstable boulder slopes with little soil development, and crustose lichens cover the rocks. Young glacial moraines in the upper alpine that have only recently been exposed can be found in several stages of plant succession.

The Himalayan alpine is home to several large ungulates, including goats, ibex, and tahr, and blue sheep or bharal. The tahr, related to wild goats, has a double coat for warmth and hooves well suited to gripping rocks. Bharal is neither sheep nor goat but is a larger mammal with characteristics of each. High meadows in both the Himalayas and Tibet are grazed by yaks. This large animal, males weighing about 1,000 lb (450 kg), survives by grazing grasses, forbs, and lichens. Like the muskox of the Arctic, the yak has a dense undercoat covered by longer guard hairs to give it insulation during cold winters. A large lung capacity and numerous red blood cells enable it to cope with low oxygen at high elevations. Yaks conserve energy when traveling over snow in winter by walking single file and carefully stepping in the footprints of the lead animal. Usually congregating in herds of 10–30, yaks were formerly more widespread. Numbers have decreased as a consequence of both uncontrolled hunting and conversion of grazing grounds to pasture for domesticated animals. The only horse native to alpine regions anywhere in the world is the Tibetan wild ass, or kiang. Smaller herbivores include marmots, pikas, hares, and voles.

Large carnivores include the elusive snow leopard, less than 4 ft (1.2 m) long but with a 3 ft (1 m) tail that provides balance in rugged terrain (see Plate XII). The snow leopard has the thickest coat of any cat and also has thick fur on the soles of its feet. Enlarged nasal cavities help it breathe thin air. Brown bears are occasionally seen, and the Himalayan black bear wanders into the alpine from the forest below. Like their Arctic or North American counterparts, both bears are omnivorous. Red fox, Tibetan wolf, and Siberian weasel are smaller predators.

Counterparts of the ptarmigan in North American mountains are Snow Partridge, Himalayan Snow Cock, and Tibetan Snow Cock, all members of the pheasant family. Raptors include Golden Eagle, Lammergeier or Bearded Vulture, and Himalayan Griffon Vulture.

Few locations in the Himalayas have not been influenced by human activity. Destruction of the upper forest belt by fire has had the most effect, allowing mat-forming, nongrazed alpine plants to extend their distribution to lower elevations. Cattle and goats graze up to the highest alpine zone, even to 17,400 ft (5,300 m) in Tibet. Secondary growth is rich in flowering plants such as species of rock primrose, gentians, iris, and cinquefoil. After grazing has opened the natural vegetation, woody weeds appear. Rhododendrons and junipers, which are not grazed, are periodically removed by fire to maintain plants suitable for forage, such as kobresia sedges. Forests on shady, humid slopes are less susceptible to fire and are removed only in patches.

## Southern Hemisphere

Gradients along slopes from exposed rocky ridges to valley bottoms are similar to those in the Northern Hemisphere, but they control a different flora. Very small dwarf shrubs are the most characteristic plants in all alpine habitats.

*Southern Africa.* Although the high plateau of Lesotho, 29° S, is a treeless zone at 11,000 ft (3,300 m), temperatures are higher than in most alpine areas (see Figure 3.16). The plateau consists of the Thaba-Putso Range, Central Range, and

## Southern Andes

The high-elevation region in central Argentina and part of Chile at 27°–39° S can be considered either alpine tundra or grassland because it contains elements of both. Elevations drop from 14,000 ft (4,250 m) in the north to about 5,500 (1,700 m) in the south. Less than 25 in (630 mm) of precipitation falls annually throughout the region. January (summer) means of 55° F (13° C) are warmer than in most alpine regions, but these high elevations are covered by snow for several months in winter. A diverse flora with characteristic genera exists, and many endemic species have adaptations to extreme dry, cold, and windy conditions with possibility of frost year-round. A treeline of tall shrubs gives way to small shrubs, needlegrass tussocks, and cushion plants. In the subnival zone, small forbs, rosettes, and small grasses dominate, but cushions are common. Conspicuous endemic shrubs are *Adesmia pinifolia*, *Chuquiraga ruscifolia*, and *C. echegarayi*. Endemic cushions include *Adesmia subterranea*, *Azorella cryptantha*, *Laretia acaulis*, and *Oxychloe bisexualis*. Although most animal life is related to Andean dry puna (see Chapter 4) and Patagonian steppe, some animals, such as the nearly extinct wild chinchilla, Patagonian chinchilla mouse, and Andean field mouse, are endemic.

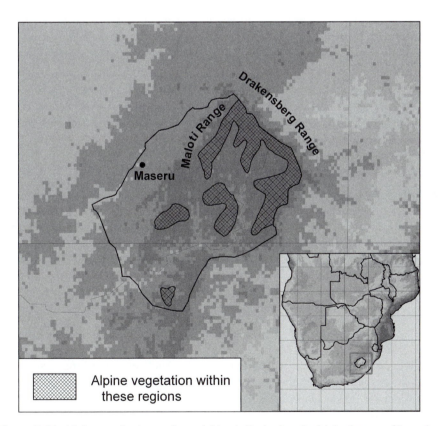

**Figure 3.16** Alpine tundra in southern Africa is limited to the high plateau of Lesotho. *(Map by Bernd Kuennecke.)*

Maloti Mountains and is bounded by the Drakensberg Mountains (the east Africa escarpment) on the east. The highest peak on the plateau is Thabana Ntlenyana (Pretty Little Mountain) at 11,430 ft (3,484 m). The plateau is broad and dissected, with boggy areas at the headwaters of the Orange River system. Geology consists of basaltic lava up to 5,250 ft (1,600 m) thick covered by thin to no soils and rock rubble. During summers, the region is wet and the ground is waterlogged. Winters bring nightly freezing and daily thawing, formation of needle ice, solifluction, and patterned ground, all characteristics of alpine or tundra landscapes.

Low temperature combined with low precipitation in winter is stressful for vegetation. Snow falls about eight times per year in winter (April to September), but usually melts quickly so there is no long-lying snow cover. Although the frost-free season is about six months, the growing season is still short. Mean temperature of the warmest month (January) is 52° F (11° C), but temperatures can be extreme. The highest temperature recorded was 88° F (31° C). The lowest was −5° F (−21° C). The climate is generally humid, and the mean annual precipitation varies from 25 in (630 mm) to 63 in (1,600 mm). Highest precipitation occurs on the eastern side, part of the Drakensberg Mountains, with rainshadow on the western side of the escarpment. Relative humidity varies 18–72%, according to the wet summer season or dry winter season. High winds in late winter and spring are especially detrimental to plants because soils are already dry at that time. Fire has been part of the ecology of the plateau, both induced by humans to increase pasture and caused by natural lightning.

Vegetation is primarily heath communities dominated by low woody species of erica shrubs and strawflower cushions scattered in extensive grasslands. Alpine heath is the dominant vegetation, with dwarf shrubs less than 24 in (60 cm) high. However, much has been cut for fuel. Boulder fields support a taller heath, 2–4 ft (0.5–1.2 m) high, with aster-like plants and strawflower.

Outcrops of horizontal basalt layers have many pioneer species of lichens and mosses. A succulent euphorbia called lion's spoor forms cushions on sunny and rocky north-facing slopes. Wiregrass tussocks and a dwarf bulrush are found in wetter areas. Three semiwoody shrubs, euryops daisy, strawflower, and a member of the milkwort family, all form cushions. Alpine grassland ranges from a low turf to tall plants in an open landscape, interrupted by mud patches. Three temperate genera of grasses—fescue, wiregrass, and *Pentaschistis*—dominate, with wiregrass being most important. All are tufted, drought tolerant, and dormant in winter. With enough water, wiregrass tussocks can be up to 3.3 ft (1 m) tall. Other grasses include tufted hairgrass and lovegrass according to aspect. Flowering forbs are numerous in spring and autumn.

Bogs develop in swales, where aquatic plants such as aponogeton, a succulent crassula, oxygen weed, mudwort, and algae grow in open pools. Pool edges may have pipewort, poker plants, bluegrass, buttercups, and bladderworts. Bog carpets without open water have hummocks of bush tea and everlasting, with many mosses and a variety of higher plants. Drier soil at the edges of the bogs are predominantly carex sedge meadows.

Due to climate and hunting, vertebrate fauna is species poor. Larger mammals such as antelope are no longer present. Major small mammals are the ice rat that digs up soil and peat in bogs, and the common mole rat. Insects, especially flies, beetles, and bees, are still numerous. The area has long been used as pasture for domestic animals and is overgrazed and degraded, with serious trampling of the bogs that constitute important headwaters.

***New Zealand.*** Most of the alpine region in New Zealand is on South Island at 41°–47° S latitude (see Figure 3.17). Treeline is roughly 3,300 ft (1,000 m), and

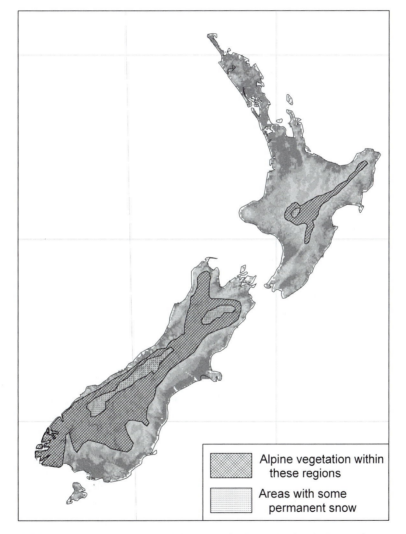

**Figure 3.17** Characteristics of the alpine zone in New Zealand change from west to east because of the southern Alps' location in the path of westerly winds. *(Map by Bernd Kuennecke.)*

permanent snow begins at 6,550 ft (2,000 m). The indigenous plants and animals of both islands are largely endemic. Human influences during the last 100 years, however, have changed alpine communities. Treeline in many places has been destroyed by burning, allowing tussock grasses to extend farther downslope. Today, the upper limit of fescue tussocks coincides with natural treeline, and climate factors—such as average summer temperatures of less than 50° F (10° C)—or indicator species, rather than vegetation, define treeline and the alpine zone. New Zealand's mountains are young, and North Island's volcanoes are still active, so lava and loose scoria are common. South Island's geology is more varied and includes igneous, metamorphic, and sedimentary rock. Extensive moraines and outwash from alpine glaciation on South Island provide parent material for unstable soils, such as poorly developed inceptisols and lithosols. Freeze-thaw cycles are active in moist soils, and small-scale patterned ground such as stone polygons, stripes, and solifluction terraces and lobes are present.

Climate is generally maritime. The westcoast location of mountains combined with prevailing westerly winds and storm tracks result in a distinct west to east climate gradient, particularly on South Island. Precipitation on the windward side is more than 160 in (4,000 mm) per year and may reach 470 in (12,000 mm) on some slopes. Lee sites receive 50 in (1,250 mm), with less than 14 in (350 mm) in the driest valleys. Gradients also exist for solar radiation, temperature, and humidity, but overall, the climate is moist with no soil moisture deficit. Even when warming from downslope winds increases temperature, cool nights raise relative humidity to saturation. The moderate temperatures reflect the maritime influence felt throughout the islands. Mean summer temperature at the upper limit of the low-alpine zone is 40° F (5° C), decreasing to 32° F (0° C) in the high alpine. Temperature extremes in the low-alpine zone are 75° F (24° C) in the summer and −2° F (−19° C) in the winter; frost is common almost nightly. Needle ice may occur on summer nights. The longest recorded frost-free period was only 13 days. Winters are much warmer than in continental mountains and rarely drop below 30° F (−1° C) in the low alpine, or below 7° F (−14° C) in the high alpine. Seasonal soil temperatures vary less than air temperatures, and even in the high-alpine soil only freezes 4 in (10 cm) deep. Winds redistribute snow cover, blowing it into drifts and insulating the soil. Summer days can be sunny even with the common occurrence of fog and low clouds. If surfaces become parched by sun during the day, they will be relieved come nightfall.

Although growing under conditions less severe than either the arctic or mid-latitude alpine zones, the alpine flora of the Southern Alps has evolved many typical tundra vegetation forms—cushions, rosettes, mats, succulents, perennials, taproots, and hairy or wooly coverings. Many plants have close relatives (that is, cogeners) in other parts of the world, while many others are native only to New Zealand or to New Zealand and a few islands off Antarctica.

One peculiarity of this alpine region is the abundance of fleshy succulent plants found particularly on sunny, dry scree slopes. Fleshy structures preserve moisture during the day, a distinct advantage since groundwater is unavailable on extremely well-drained talus.

The most distinctive plants of the New Zealand alpine flora are two endemic genera called vegetable sheep—so named because, at a distance, a group of them resembles a flock of sheep. The genera, *Haastii* and *Raoulia*, are both members of the sunflower family, but they are not closely related. *Haastii* forms hummocks of either tight rosettes or loose branches, covered with fine wooly or wispy hairs. *Haastii* are found in dry mountain areas where in summer sunshine alternates with cold, foggy winds. By experimentation, it was found that the hairs prevent water loss via excessive transpiration and condense every possible bit of moisture in the air during droughts to keep the interior tissues moist. Two species of *Raoulia* consist of a hard, rounded surface that makes a strong crust around peaty humus. The surface is covered with fine white hairs that shed water easily. Because the purpose of the hairs is to shed excess moisture rather than absorb the maximum air moisture possible, *Raoulia* are adapted to and found in wet mountain areas. This use for plant hairs may be unique to the vegetable sheep of New Zealand.

The maritime climate allows a long growing season compared with most other alpine areas. For snow-tussock grasses, vegetative growth takes place for about eight months at the lower limit of the alpine zone and only decreases to five months in the high alpine. High-alpine cushion plants produce flowers over an eight-month time span (October to May), although different species bloom at different times. Despite the long growing season, most flower buds preform the season prior to blooming.

Frequently no krummholz forms at treeline in New Zealand, and the change from forest to alpine tundra is abrupt. Erect southern beech trees give way to low shrubs of various species. These low shrubs are not environmentally stunted forms of forest species, but are genetically short. Only under severe conditions of snow and winter desiccation does treeline beech grow as krummholz, with semiprostrate stems up to 10 ft (3 m) and sometimes forming flag trees.

The low-alpine zone, a narrow elevational belt, is dominated by tall evergreen snow tussock grasses, shrubs, and forbs. Grass tussocks grow as tall as 4 ft (1.2 m), unusual in alpine regions outside the Tropics, but probably made possible by the mild climate. Large forbs and shrubs are conspicuous on the wetter western mountains, especially on rocky ridges and sunny aspects where snow fails to accumulate. A variety of plants sort themselves according to soil and microhabitat conditions, especially snow cover. Evergreen shrubs might be *Dracophyllum* and *Olearia;* common herbaceous plants are alpine daisy and New Zealand flax. Most widespread in high rain areas, bogs with underlying peat occupy poorly drained depressions with high water tables. Plants include several cushion plants such as alpine daisy along with *Oreobolus* sedges. Several low, trailing subshrubs and herbs grow in bogs, where insectivorous sundews, bladderworts, fruticose lichens, sedges, and rushes are also common.

The high-alpine zone has a discontinuous cover of shorter grasses, forbs, and shrubs, with several distinct communities. Fellfields of stable rock are sparsely vegetated. The drier areas are characterized by cushion plants, particularly vegetable sheep, and small *Hebe* shrubs, strawflower, and cushion speargrass. Wet fellfields have more diversity, with more small herbs and grasses. Since talus or scree of loose, angular stones on steep slopes can be mobile, especially during spring snowmelt, such areas support only a sparse cover of summer-green herbs. Plants such as fireweed and buttercup have strong roots penetrating into the deep soils beneath the rocks that enable them to withstand scree movement.

Cushion fields on broad plateau summits with strong winds, cold summers, and freeze-thaw cycles have dwarf cushions, mats, and prostrate plants less than 0.75 in (2 cm) tall, vegetation similar to that found in arctic tundra. Where abraded or scoured by strong wind, they may assume asymmetrical shapes. *Dracophyllum muscoides* is a prevalent prostrate shrub. Patterned ground contributes to a mosaic on the landscape. Snowbanks, which persist until mid or late summer in depressions and areas sheltered from prevailing winds, are generally well vegetated. The smallest snow tussock grass is dominant, forming dense turf 4 in (10 cm) high, but many other plants grow in this microenvironment as well.

The nival zone, with lichens on rock faces too steep to maintain snow cover, is extensive in the Southern Alps. Three flowering plants grow above 8,200 ft (2,500 m) on Mt. Cook, a buttercup and subshrubs *Hebe haastii* and *Parahebe birleyi*.

Except for two species of bat, no mammals are indigenous to the alpine zone. Birds, however, are represented by many indigenous and endemic breeding species. Several are rare and endangered. A distinctively alpine bird is the Flightless Rail or Takahe, now limited to Fiordland National Park. It feeds in low-alpine tussock grasslands, taking the tillers of the plants. Today, the introduced red deer offers serious competition. The endemic Alpine Parrot, the Kea, is widespread and common in the Southern Alps, but spends winters below treeline (see Figure 3.18). This noisy, olive-green and red bird is omnivorous and, in spite of being only 12 in (30 cm) tall, has successfully attacked and killed sheep. The widespread New Zealand Falcon is a strict carnivore. The larger raptor, Australasian Harrier, soars on thermals over grasslands, occasionally rising as high as the alpine zone. The only bird restricted to the alpine environment of South Island is the small Rock Wren, which feeds on insects and fruit and spends the winter in rock crevices beneath the snow. Some coastal birds, such as the South Island Pied Oystercatcher and Banded Dotterel, migrate to the alpine for breeding and nesting in cushion fields. The Southern Black-backed Gull, as well as other gulls, nest in bogs. The New Zealand Pipit may also nest in alpine zones. Few seed eaters are found in the alpine zone because the seed supply is unreliable. Reptiles are few; two skinks and two geckos are known to live at high elevations.

Several vertebrates have been introduced. Red deer are widely distributed throughout the mountains and are most damaging by virtue of their heavy grazing and trampling of the soil. Widespread hares also detrimentally affect the alpine zone. Goats, wapiti, chamois, and tahr are more limited in extent but do damage where they occur.

**Figure 3.18** The Kea is a alpine parrot native to New Zealand. *(Courtesy of Shutterstock. Copyright: Dario Diament.)*

Invertebrate fauna is rich in all habitats free of snow in summer up to 9,800 ft (3,000 m). Insects are important contributors to the alpine ecosystem, pollinating some plants, defoliating others, and consuming litter. Conspicuous in size and number of species are wetas, grasshoppers, beetles, moths, butterflies, cicadas, and flies, especially black flies and blow flies. Characteristic of alpine New Zealand are several genera of giant weta (Orthoptera: Stenopelmatidae), large beetles that live under rock slabs. With bodies up to 2.75 in (70 mm long), the insect weighs up to 2.5 oz (70 g). More than 40% of New Zealand's moths and butterflies are alpine in distribution; some are endemic. Several diurnal wolf spiders and dark-colored jumping spiders overwinter beneath the snow. They are most diverse on South Island. Because of the short growing season, insects have a short active period. It may take more than one year to complete the generation cycle, with larvae waiting out unfavorable periods concealed by host plants. Common adaptations include dark color, long and dark body hair, and a large body.

## Further Readings

Blumstein, Daniel T. n.d. "Marmot Lore 101: Class Notes." Marmotburrow. http://www.marmotburrow.ucla.edu/lore.html.

Bowen, Ezra. 1972. *The High Sierra, The American Wilderness*. New York: Time-Life.

European Environmental Agency. n.d. "Mountain Distribution of Plants." Epaedia, Environment Explained. http://epaedia.eea.europa.eu/page.php?pid=341.

Great Himalayan National Park. n.d. Great Himalayan National Park. http://www.greathimalayannationalpark.com/GHNP_home.htm.

Price, Larry W. 1981. *Mountains and Man: A Study of Process and Environment.* Berkeley: University of California Press.

Ross, Graham. n.d. The Alpine Plants of New Zealand. http://alpine-plants-new-zealand.110mb.com.

Zwinger, Ann H., and Beatrice E. Willard. 1972. *Land Above the Trees: A Guide to American Alpine Tundra.* New York: Harper & Row.

# Appendix

## Biota of the Mid-Latitude Alpine Tundra Biome
(arranged geographically)


### North American Alpine Tundra

#### Some Characteristic Plants of the Coastal Ranges

*Treeline and krummholz*

| | |
|---|---|
| Mountain hemlock | *Tsuga mertensiana* |
| Subalpine fir | *Abies lasiocarpa* |
| Alaska cedar | *Chamaecyparis nootkatensis* |

*Low shrubs or subshrubs*

| | |
|---|---|
| Resin birch | *Betula glandulosa* |
| Tea-leaved willow | *Salix planifolia* |
| Arctic willow | *Salix arctica* |
| Black crowberry | *Empetrum nigrum* |
| Mountain heather | *Phyllodoce* spp. |

*Graminoids*

| | |
|---|---|
| Mountain hare sedge | *Carex phaeocephala* |
| Altai fescue grass | *Festuca altaica* |

*Forbs*

*Cushions or mats*

| | |
|---|---|
| Moss campion | *Silene acaulis* |
| Tufted saxifrage | *Saxifraga caespitosa* |
| Spreading phlox | *Phlox diffusa* |
| Creeping sibbaldia | *Sibbaldia procumbens* |

*Leafy or rosettes*

| | |
|---|---|
| Alpine bittercress | *Cardamine bellidifolia* |
| Chiming bells | *Mertensia paniculata* |
| Lousewort | *Pedicularis* spp. |
| Fireweed | *Epilobium* spp. |

*Cryptogams*

| | |
|---|---|
| Jewel lichen (Crustose) | *Caloplaca elegans* |
| Sphagnum moss | *Sphagnum* ssp. |

## Some Characteristic Animals of the Coastal Ranges

*Herbivores*

| | |
|---|---|
| Mountain goat | *Oreamnos americana* (Introduced) |
| Dall's sheep | *Ovis dalli* (Alaska and Canada only) |
| Olympic marmot | *Marmota olympus* (Endemic to Olympics) |
| Hoary marmot | *Marmota caligata* (Alaska and Canada only) |
| American pika | *Ochotona princeps* |
| Mazama pocket gopher | *Thomomys mazama* |
| Deer mouse | *Peromyscus maniculatus* |
| Meadow vole | *Microtus pennsylvanicus* |

*Carnivores*

| | |
|---|---|
| Short-tailed weasel or Ermine | *Mustela erminea* |
| Dusky shrew | *Sorex monticolus* |
| Brown or Grizzly bear | *Ursus arctos* |

*Birds*

| | |
|---|---|
| White-tailed Ptarmigan | *Lagopus leucurus* |
| Horned Lark | *Eremophila alpestris* |
| Water Pipet | *Anthus spinoletta* |
| Gray-crowned Rosy Finch | *Leucosticte arctoa tephrocotis* |
| Golden Eagle | *Aquila chrysaetos* |
| Red-tailed Hawk | *Buteo jamaicensis* |
| Stellar's Jay | *Cyanocitta stelleri* |

## Some Characteristic Plants of the Cascade Range

*Treeline and krummholz*

| | |
|---|---|
| Whitebark pine | *Pinus albicaula* |
| Subalpine fir | *Abies lasiocarpa* |
| Engelmann spruce | *Picea engelmannii* |
| Mountain hemlock | *Tsuga mertensiana* |
| Alpine larch | *Larix lyallii* |

*Low shrubs or subshrubs*

| | |
|---|---|
| Black crowberry | *Empetrum nigrum* |
| Partridge foot | *Luetkea pectinata* |
| Kinnikinnik or Bear-berry | *Arctostaphylos uva-ursi* |
| Western moss heather | *Cassiope mertensiana* |
| Pink mountain heather | *Phyllodoce empetriformis* |

### Graminoids

| | |
|---|---|
| Mountain hairgrass | *Deschampsia atropurpurea* |
| Kobresia sedge | *Kobresia myosuroides* |
| Blackish sedge | *Carex nigricans* |

### Forbs

*Cushions or mats*

| | |
|---|---|
| Tolmie saxifrage | *Saxifraga tolmiei* |

*Leafy or rosettes*

| | |
|---|---|
| Cliff Indian paintbrush | *Castilleja rupicola* |
| Lemmon's rockcress | *Arabis lemmonii* |
| Lyall lupine | *Lupinus lyallii* |
| Alpine aster | *Aster alpigenus* |
| Bistort | *Polygonum bistortoides* |
| Newberry knotweed | *Polygonum newberryi* |
| Pussypaws | *Calyptridium umbellatum* |

### Cryptogams

| | |
|---|---|
| Jewell lichen (Crustose) | *Caloplaca elegans* |

## Some Characteristic Animals of the Cascade Range

### Herbivores

| | |
|---|---|
| Mountain goat | *Oreamnos americana* (Introduced) |
| Yellow-bellied marmot | *Marmota flaviventris* |
| American pika | *Ochotona princeps* |
| Deer mouse | *Peromyscus maniculatus* |
| Heather vole | *Phenacomys intermedius* |
| Mazama pocket gopher | *Thomomys mazama* |
| American water vole | *Microtus richardsonii* |

### Carnivores

| | |
|---|---|
| Short-tailed weasel or Ermine | *Mustela erminea* |
| Dusky shrew | *Sorex monticolus* |

### Birds

| | |
|---|---|
| Horned Lark | *Eremophila alpestris* |
| Gray-crowned Rosy Finch | *Leucosticte arctoa tephrocotis* |
| Rock Wren | *Salpinctes obsoletus* |
| Water Pipet | *Anthus spinoletta* |
| Golden Eagle | *Aquila chrysaetos* |
| Red-tailed Hawk | *Buteo jamaicensis* |
| Clark's Nutcracker | *Nucifraga columbiana* |
| Stellar's Jay | *Cyanocitta stelleri* |
| White-tailed Ptarmigan | *Ptarmigan leucurus* |

## Some Characteristic Plants of the Sierra Nevada

*Timberline and krummholz*

Whitebark pine                          *Pinus albicaulis*
Mountain hemlock                        *Tsuga mertensiana*
Sierra lodgepole pine                   *Pinus contorta* var. *murrayana*
Foxtail pine                            *Pinus balfouriana*

*Low shrubs or subshrubs*

Dwarf bilberry                          *Vaccinium caespitosum*
Timberline sagebrush                    *Artemisia rothrockii*

*Graminoids*

Shorthair reedgrass                     *Calamagrostis breweri*
Short-grass sedge                       *Carex exserta*

*Forbs*

*Cushions or mats*
Sierra saxifrage                        *Saxifraga aprica*
Alpine pussytoes                        *Antennaria alpina*
Clubmoss ivesia                         *Ivesia lycopodioides*
Tufted phlox                            *Phlox caespitosa*
Davidson's penstemon                    *Penstemon davidsonii*

*Leafy or rosettes*
Nutall's sandwort                       *Arenaria nuttallii*
Dwarf knotweed                          *Polygonum minimum*
Alpine sorrel                           *Oxyria digyna*
Bottlebrush squirreltail                *Sitanion hystrix*
Great Basin violet                      *Viola beckwithii*
Oval-leaved buckwheat                   *Eriogonum ovalifolium*
Prickly gilia                           *Leptodactylon pungens*
Pussypaws                               *Calyptridium umbellatum*

*Cryptogams*

Jewell lichen (Crustose)                *Caloplaca elegans*

## Some Characteristic Animals of the Sierra Nevada

*Herbivores*

American pika                           *Ochotona princeps*
Yellow-bellied marmot                   *Marmota flaviventris*
Alpine chipmunk                         *Tamias alpinus* (Endemic)
Sierra pocket gopher                    *Thomomys monticola*
Deer mouse                              *Peromyscus maniculatus*
Heather vole                            *Phenacomys intermedius*
Bighorn sheep                           *Ovis canadensis*

### Carnivores

Short-tailed weasel or Ermine | *Mustela erminea*
Dusky shrew | *Sorex monticolus*

### Birds

White-tailed Ptarmigan | *Lagopus leucurus* (Introduced)
Rosy Finch | *Leucosticte arctoa*
Rock Wren | *Salpinctes obsoletus*
Horned Lark | *Eremophila alpestris*
Clark's Nutcracker | *Nucifraga columbiana*
Golden Eagle | *Aquila chrysaetos*
Red-tailed Hawk | *Buteo jamaicensis*
Stellar's Jay | *Cyanocitta stelleri*

## Some Characteristic Plants of the Great Basin

### Treeline and krummholz

Whitebark pine | *Pinus albicaulis*
Engelmann spruce | *Picea engelmannii*
Subalpine fir | *Abies lasiocarpa*
Limber pine | *Pinus contorta*
Big sagebrush | *Artemisia tridentata*
Bristlecone pine | *Pinus longaeva*

### Low shrubs or subshrubs

Arctic willow | *Salix arctica*

### Graminoids

Alpine reedgrass | *Calamagrostis purpurascens*
Alpine fescue grass | *Festuca brachyphylla*
Spiked trisetum grass | *Trisetum spicatum*
Sheep fescue grass | *Festuca ovina*
Mountain hare sedge | *Carex phaeocephala*
Elk sedge | *Carex elynoides*
Rocky Mountain sedge | *Carex scopulorum*

### Forbs

*Cushions or mats*
Coville's phlox | *Phlox covillei*
Alpine avens | *Geum rossii*
Moss campion | *Silene acaulis*

*Leafy or rosettes*
Mono clover | *Trifolium monoense*
Bottlebrush squirreltail | *Sitanion hystrix*
Broad-podded phoenicaulis | *Phoenicaulis eurycarpa*

*(Continued)*

| | |
|---|---|
| Mason's Jacob's ladder | *Polemonium chartaceum* |
| Rambling fleabane | *Erigeron vagus* |
| Marsh marigold | *Caltha leptosepala* |
| American bistort | *Polygonum bistortoides* |

**Cryptogams**

| | |
|---|---|
| Jewell lichen (Crustose) | *Caloplaca elegans* |

## Some Characteristic Animals of the Great Basin

**Herbivores**

| | |
|---|---|
| Yellow-bellied marmot | *Marmota flaviventris* |
| Bighorn sheep | *Ovis canadensis* (Rare) |
| American pika | *Ochotona princeps* |
| Deer mouse | *Peromyscus maniculatus* |
| Northern pocket gopher | *Thomomys talpoides* |

**Carnivores**

| | |
|---|---|
| Short-tailed weasel or Ermine | *Mustela erminea* |

**Birds**

| | |
|---|---|
| Rock Wren | *Salpinctes obsoletus* |
| Horned Lark | *Eremophila alpestris* |
| Golden Eagle | *Aquila chrysaetos* |
| Red-tailed Hawk | *Buteo jamaicensis* |
| Clark's Nutcracker | *Nucifraga columbiana* |

## Some Characteristic Plants of the Rocky Mountains

**Krummholz and treeline**

| | |
|---|---|
| Engelmann spruce | *Picea engelmannii* |
| Subalpine fir | *Abies lasiocarpa* |
| Whitebark pine | *Pinus albicaulis* |
| Limber pine | *Pinus contorta* |
| Alpine larch | *Larix lyallii*[a] |
| Tamarack | *Larix laricina*[a] |
| Western hemlock | *Tsuga heterophylla*[a] |
| White spruce | *Picea glauca*[a] |
| Black spruce | *Picea mariana*[a] |

**Low shrubs or subshrubs**

| | |
|---|---|
| Arctic willow | *Salix arctica* |
| Downy birch | *Betula pubescens*[a] |
| Sibbaldia | *Sibbaldia procumbens* |
| Arctic bell heather | *Cassiope tetragona*[a] |
| Black crowberry | *Empetrum nigrum*[a] |
| Mountain heather | *Phylladoce* spp.[a] |

| | |
|---|---|
| Marsh Labrador tea | *Ledum palustre*[a] |
| Bog bilberry | *Vaccinium uliginosum*[a] |

### Graminoids

| | |
|---|---|
| Tufted hairgrass | *Deschampsia caespitosa* |
| Kobresia sedge | *Kobresia myosuroides* |
| Elk sedge | *Carex elynoides* |
| Cottongrass sedge | *Eriophorum angustifolium* |
| Mountain hare sedge | *Carex phaeocephala*[a] |
| Bigelow's sedge | *Carex bigelowii*[a] |
| Rocky Mountain sedge | *Carex scopulorum* |
| Drummond's rush | *Juncus drummondii* |
| Northern woodrush | *Luzula confusa*[a] |

### Forbs

*Cushions or mats*

| | |
|---|---|
| Saxifrage | *Saxifraga* spp. |
| Alpine phlox | *Phlox condensata* |
| Alpine candytuft | *Thlaspi alpestre* |
| Moss campion | *Silene acaulis* |
| Dwarf clover | *Trifolium nanum* |
| One-flowered harebell | *Campanula uniflora* |
| American bistort | *Polygonum bistortoides* |
| Sky pilot | *Polemonium viscosum* |
| Alpine avens | *Geum rossii* |
| Alpine sage | *Artemisia borealis* |
| Alpine locoweed | *Oxytropis podocarpa*[a] |
| Dotted saxifrage | *Saxifraga bronchialis*[a] |
| Mountain avens | *Dryas octopetala* |

*Leafy or rosettes*

| | |
|---|---|
| Alpine sorrel | *Oxyria digyna* |
| Alpine primrose | *Primula angustifolia* |
| Milkvetch | *Astralagus plumbeus* |
| Draba | *Draba* spp. |
| Ragwort | *Senecio* spp. |
| Arctic gentian | *Gentiana algida* |
| Snow buttercup | *Ranunculus adoneus* |
| King's crown | *Sedum rosea* |
| Viviparous bistort | *Polygonum viviparum* |
| Marsh marigold | *Caltha leptosepala* |

### Cryptogams

| | |
|---|---|
| Iceland lichen (Fruticose) | *Cetraria islandica* |
| Snow lichen (Fruticose) | *Cetraria nivalis* |

*(Continued)*

| | |
|---|---|
| Knob lichen (Fruticose) | *Dactylina madreporiformis* |
| Worm lichen (Fruticose) | *Thamnolia vermicularis* |
| Reindeer lichen (Fruticose) | *Cladonia rangifera*[a] |
| Jewell lichen (Crustose) | *Caloplaca elegans* |
| Rock clubmoss | *Selaginella densa*[a] |

*Note:* [a]Northern Rocky Mountains, Alaska, and Yukon only.

## Some Characteristic Animals of the Rocky Mountains

### *Herbivores*

| | |
|---|---|
| Elk | *Cervus elaphus* |
| Mountain bighorn sheep | *Ovis canadensis*[b] |
| Dall's sheep | *Ovis dallii*[a] |
| Mountain goat | *Oreamnos americana*[a] |
| Deer mouse | *Peromyscus maniculatus* |
| Northern pocket gopher | *Thomomys talpoides*[b] |
| Heather vole | *Phenacomys intermedius* |
| American pika | *Ochotona princeps*[b] |
| Collared pika | *Ochotona collaris*[a] |
| Yellow-bellied marmot | *Marmota flaviventris*[b] |
| Hoary marmot | *Marmota caligata*[a] |
| Alaska marmot | *Marmota broweri*[a] |
| Water vole | *Microtus richardsoni* |
| Meadow vole | *Microtus pennsylvanicus* |

### *Carnivores*

| | |
|---|---|
| Brown or Grizzly bear | *Ursos arctos*[a] |
| Short-tailed weasel or Ermine | *Mustela erminea* |
| Dusky shrew | *Sorex monticolus* |

### *Birds*

| | |
|---|---|
| White-tailed Ptarmigan | *Lagopus leucurus* |
| Water Pipit | *Anthus spinoletta* |
| Rosy Finch | *Leucosticte arctoa* |
| Rock Wren | *Salpinctes obsoletus*[b] |
| Horned Lark | *Eremophila alpestris* |
| Stellar's Jay | *Cyanocitta stelleri* |
| Clark's Nutcracker | *Nucifraga columbiana* |
| Golden Eagle | *Aquila chrysaetos* |
| Red-tailed Hawk | *Buteo jamaicensis* |

*Notes:* [a]Northern Rocky Mountains, Alaska, and Yukon only; [b]Southern and Central Rocky Mountains only.

## Some Characteristic Plants of Mount Washington

### Treeline and krummholz
| | |
|---|---|
| Balsam fir | *Abies balsamea* |
| Black spruce | *Picea mariana* |
| Paper birch | *Betula papyrifera* |

### Low shrubs or subshrubs
| | |
|---|---|
| Bog bilberry | *Vaccinium uliginosum* |
| Mountain heather | *Phyllodoce caerulea* |
| Mountain cranberry | *Vaccinium vitis-idaea* |
| Sweet blueberry | *Vaccinium angustifolium* |
| Three-toothed cinquefoil or Sibbaldiopsis | *Potentilla tridentata* |
| Bearberry willow | *Salix uva-ursi* |
| Mountain alder | *Alnus biridus* ssp. *crispa* |
| Labrador tea | *Ledum groenlandicum* |

### Graminoids
| | |
|---|---|
| Nodding hairgrass | *Deschampsia flexuosa* |
| Bigelow's sedge | *Carex bigelowii* |
| Highland rush | *Juncus trifidus* |

### Forbs
*Cushions or mats*
| | |
|---|---|
| Greenland sandwort | *Arenaria groenlandica* |
| Lapland diapensia | *Diapensia lapponica* |

*Leafy or rosettes*
| | |
|---|---|
| Dwarf willow | *Salix herbaceae* |
| Viviparous bistort | *Polygonum viviparum* |

### Cryptogams
| | |
|---|---|
| Iceland lichen (Fruticose) | *Cetraria islandica* |
| Haircap moss | *Polytrichum juniperinum* |
| Clubmoss | *Lycopodium* spp. |

## Some Characteristic Animals of Mount Washington

### Herbivores
| | |
|---|---|
| Groundhog | *Marmota monax* |
| Deer mouse | *Peromyscus maniculatus* |
| Meadow vole | *Microtus pennsylvanicus* |

### Carnivores
| | |
|---|---|
| Short-tailed weasel or Ermine | *Mustela erminea* |
| Masked shrew | *Sorex cinereus* |

*Birds*

| | |
|---|---|
| Slate-colored Junco | *Junco hyemalis* |
| White-throated Sparrow | *Zonotrichia albicaulis* |
| Red-tailed Hawk | *Buteo jamaicensis* |

## Eurasian Alpine Tundra

## Some Characteristic Plants of the European Alps

*Treeline*

| | |
|---|---|
| Norway spruce | *Picea abies* |
| Prostrate pine | *Pinus mugo* |
| European larch | *Larix decidua* |
| Stone pine | *Pinus cembra* |
| Mountain pine | *Pinus uncinata* |

*Low shrubs or subshrubs*

| | |
|---|---|
| Lapland rosebay | *Rhododendron lapponicum* |
| Alpenrose | *Rhododendron ferrugineum* |
| Bog bilberry | *Vaccinium uliginosum* |
| Creeping azalea | *Loiseleuria procumbens* |
| Green alder | *Alnus viridus* |
| Net-leaved willow | *Salix reticulata* |

*Graminoids*

| | |
|---|---|
| Matgrass | *Nardus stricta* |
| Fescue grass | *Festuca hallari* |
| Fescue grass | *Festuca varia* |
| Curly sedge | *Carex rupestris.* |
| Scorched alpine sedge | *Carex atrofusca* |
| Kobresia sedge | *Kobresia myosuroides* |

*Forbs*

*Cushions or mats*

| | |
|---|---|
| Purple saxifrage | *Saxifraga oppositifolia* |
| Moss campion | *Silene acaulis* |
| Rock jasmine | *Androsace helvetica* |
| Two-flowered saxifrage | *Saxifraga biflora* |

*Leafy or rosettes*

| | |
|---|---|
| Dwarf willow | *Salix herbaceae* |
| Glacier buttercup | *Ranunculus glacialis* |
| Edelweiss | *Leontopodium brachyactis* |
| Houseleek or Liveforever | *Sempervivum montanum* |

*Cryptogams*

| | |
|---|---|
| Norwegian haircap moss | *Polytrichum norvegicum* |

## Some Characteristic Animals of the European Alps

### Herbivores
| | |
|---|---|
| Ibex | *Capra ibex* |
| Red deer | *Cervus elaphus* |
| Chamois | *Rupricapra rupricapra* |
| Snow vole | *Arvicola nivalis* |
| Alpine marmot | *Marmota marmota* |

### Carnivores
| | |
|---|---|
| Lynx | *Lynx lynx*[a] |
| Brown bear | *Ursus arctos*[a] |
| Wildcat | *Felis silvestris* |
| Alpine shrew | *Sorex alpinus* |

### Birds
| | |
|---|---|
| Lammergeier or Bearded<br>  Vulture | *Gypaetus barbatus*[a] |
| Golden Eagle | *Aquila chrysaetos*[a] |

*Note:* [a]Extinct or very rare.

## Some Characteristic Plants of the Tien Shan Mountains

### Treeline
| | |
|---|---|
| Central Asian spruce | *Picea schrenkiana* |

### Shrubs
| | |
|---|---|
| Sagebrush | *Artemisia rhodantha* |
| In the Chenopod family | *Kalidium schrenkianum* |
| Shagspine pea-shrub | *Caragana jubata* |

### Graminoids
| | |
|---|---|
| Fescue grass | *Festuca sulcata* |
| False needlegrass | *Ptilagrostis subsessiliflora* |
| Needlegrass | *Stipa caucasica* |
| Spike fescue grass | *Leucopoa olgae* |
| Reedgrass | *Calamagrostis tianshanicus* |
| Tien Shan fescue | *Festuca tianshanica* |
| Alpine bluegrass | *Poa alpina* |
| Kobresia sedge | *Kobresia capilliformis* |
| Carex sedge | *Carex melanantha* |

### Forbs
*Cushions and mats*
| | |
|---|---|
| Rose family | *Dryadanthe tetranda* |
| Androsace | *Androsace sericea* |

*Leafy or rosettes*

| | |
|---|---|
| Sea lavender | *Limonium hoeltzeri* |
| Snow lotus | *Saussurea leucophylla* |
| Locoweed | *Oxytropis chionobaia* |
| Aster | *Aster* spp. |
| Gentian | *Gentiana* spp. |
| Dandelion | *Taraxacum* spp. |
| In the Buttercup family | *Callianthemum alatavicum* |
| Lousewort | *Pedicularis rhinanthoides* |
| Viviparous bistort | *Polygonum viviparum* |
| Buttercup | *Ranunculus albertii* |

**Cryptogams**

| | |
|---|---|
| Cetraria (Fruticose lichen) | *Cetraria* spp. |
| Parmelia (Foliose lichen) | *Parmelia* spp. |
| Aspicilia (Crustose lichen) | *Aspicilia* spp. |
| Twisted moss | *Tortula ruralis* |
| Wideleaf stegonia moss | *Stegonia latifolia* |

## Some Characteristic Animals of the Tien Shan Mountains

**Herbivores**

| | |
|---|---|
| Siberian ibex | *Capra sibirica* |
| Argali or Kyzylkum sheep | *Ovis ammon* |
| Gray marmot | *Marmota baibacina* |
| Tolei hare | *Lepus tolei* |
| Gray hamster | *Cricetulus migratorius* |
| Silvery mountain vole | *Alticola argentata* |
| Large-eared pika | *Ochotona macrotis* |
| Southern mountain vole | *Ellobius talpinus* |

**Carnivores**

| | |
|---|---|
| Wolf | *Canis lupus* |
| Brown bear | *Ursus arctos* |
| Red fox | *Vulpes vulpes* |
| Steppe polecat | *Mustela eversmannii* |
| Short-tailed weasel or Ermine | *Mustela erminea* |

**Birds**

| | |
|---|---|
| Horned Lark | *Eremophila alpestris* |
| Lesser Sand Plover | *Charadrius mongolus* |
| Isabelline Wheatear | *Oenanthe isabellina* |
| Snow Finch | *Montifrigilla nivalis* |
| Alpine Chough | *Pyrrhocorax graculus* |
| White-winged Redstart | *Phoenicurus erythrogaster* |
| Water Pipet | *Anthus spinoletta* |

## Some Characteristic Plants of the Himalayas

### Treeline and krummholz
| | |
|---|---|
| Western Himalayan fir | *Abies pindrow* |
| Western Himalayan spruce | *Picea smithiana* |
| Himalayan pine | *Pinus wallichiana* |
| Eastern Himalayan fir | *Abies spectabilis* |
| Drooping juniper | *Juniperus recurva* |

### Shrubs
| | |
|---|---|
| Western Himalayan birch | *Betula utilis* |
| Sea wormwood | *Artemisia maritima* |
| Rhododendron | *Rhododendron* spp. |
| Meyer single-seed juniper | *Juniperus squamata* |
| Common juniper | *Juniperus communis* |
| Grecian juniper | *Juniperus excelsa* |

### Graminoids
| | |
|---|---|
| Needlegrass | *Stipa purpurea* |
| Mount Everest sedge | *Carex montis-everestii* |
| Nepal kobresia sedge | *Kobresia nepalensis* |

### Forbs
*Cushions or mats*
| | |
|---|---|
| Snow lotus | *Saussurea gnaphalodes* |
| Sandwort | *Arenaria bryophylla* |
| Everlasting | *Anaphalis triplinervis* |
| Edelweiss | *Leontopodium brachyactis* |
| Sibbaldia | *Sibbaldia purpurea* |
| Rock primrose or Jasmine | *Androsace lehmannii* |
| Locoweed | *Oxytropis* spp. |

*Leafy or rosettes*
| | |
|---|---|
| Viviparous bistort | *Polygonum viviparum* |
| Fireweed | *Epilobium angustifolium* |
| Iceland purslane | *Koeniga islandica* |
| Alpine sorrel | *Oxyria digynia* |
| Monkshood | *Aconitum* spp. |
| Lousewort | *Pedicularis* spp. |
| Buttercup | *Ranunculus* spp. |

### Cryptogams
| | |
|---|---|
| Iceland lichen (Fruticose) | *Cetraria islandica* |
| Reindeer lichen (Fruticose) | *Cladonia rangiferina* |
| White-worm lichen (Fruticose) | *Thamnolia vermicularis* |
| Umbilicaria (Foliose lichen) | *Umbilicaria indica* |
| Ochrolechia (Crustose lichen) | *Ochrolechia glacialis* |

## Some Characteristic Animals of the Himalayas

### Herbivores

| | |
|---|---|
| Blue sheep or Bharal | *Pseudois nayur* |
| Yak | *Bos grunniens* |
| Himalayan tahr | *Hemitragus jemlahicus* |
| Ibex | *Capra ibex* |
| Tibetan wild ass or Kiang | *Equus kiang* |
| Himalayan marmot | *Marmota himalayana* |
| Himalayan pika | *Ochotona himalayana* |
| Silvery mountain vole | *Alticola argentata* |
| Tolai hare | *Lepus tolai* |

### Carnivores

| | |
|---|---|
| Snow leopard | *Panthera uncia* |
| Brown bear | *Ursus arctos* |
| Himalayan black bear | *Ursus thibetanus* |
| Red fox | *Vulpes vulpes* |
| Tibetan wolf | *Canis lupus chanku* |
| Siberian weasel | *Mustela sibirica* |

### Birds

| | |
|---|---|
| Snow Partridge | *Lerwa lerwa* |
| Himalayan Snow Cock | *Tetraogallus himalayensis* |
| Tibetan Snow Cock | *Tetraogallus tibetanus* |
| Golden Eagle | *Aquila chrysaetos* |
| Lammergeier or Bearded Vulture | *Gyptaetus barbatus* |
| Himalayan Griffon Vulture | *Gyps himalayensis* |

## Southern Hemisphere Alpine Tundra

## Some Characteristic Plants of the Lesotho Plateau

### Shrubs

| | |
|---|---|
| Erica heath | *Erica dominans* |
| Erica heath | *Erica glaphyra* |
| Strawflower or Everlasting | *Helichrysum trilineatum* |

### Graminoids

| | |
|---|---|
| Goat fescue grass | *Festuca caprina* |
| Wiregrass | *Merxmuellera drakensbergensis* |
| Copper wiregrass | *Merxmuellera disticha* |
| In the Bluegrass family | *Pentaschistis oreadoxa* |
| Bog bluegrass | *Poa binata* |
| Tufted hairgrass | *Deschampsia caespitosa* |
| Lovegrass | *Eragrostis caesia* |
| Carex sedge | *Carex* spp. |
| Dwarf bulrush | *Scirpus ficinioides* |

*Aquatic*

| | |
|---|---|
| Aponogeton | *Aponogeton junceus* |
| Crassula | *Crassula inanis* |
| Oxygen weed | *Lagarosiphon muscoidas* |
| Cape mudwort | *Limosella capensis* |

*Forbs*

*Cushions or mats*

| | |
|---|---|
| Euryops daisy | *Euryops decumbens* |
| In the Milkwort family | *Muraltia saxicola* |
| Lion's spoor | *Euphorbia clavarioides* |

*Leafy or rosettes*

| | |
|---|---|
| In the Aster family | *Eumorphia sericea* |
| Bush tea | *Athrixia fontana* |
| Poker plants | *Kniphofia caulescens* |
| Buttercup | *Ranunculus meyeri* |
| Pipewort | *Eriocaulon dregei* |
| Bladderwort | *Utricularia* spp. |

## Some Characteristic Animals of the Lesotho Plateau

*Herbivores*

| | |
|---|---|
| Ice rat | *Otomys sloggetti* |
| Mole rat | *Cryptomys hottentotus* |

## Some Characteristic Plants of the Southern Alps, New Zealand

*Treeline*

| | |
|---|---|
| Southern beech | *Nothofagus menziesii* |
| Southern beech | *Nothofagus solandri* |

*Shrubs*

| | |
|---|---|
| Vegetable sheep | *Haastii* spp. |
| Vegetable sheep | *Raoulia* spp. |
| In the Aster family | *Olearia* spp. |
| In the Figwort family | *Hebe* spp. |
| Strawflower or Everlasting | *Helichrysum* spp. |
| In the Heath family | *Dracophyllum muscoides* |
| In the Figwort family | *Parahebe birleyi* |

*Graminoids*

| | |
|---|---|
| Red snow tussock grass | *Chionochloa rubra* |
| Snow tussock grass | *Chionochloa oreophila* |
| Bentgrass | *Agrostis* spp. |
| Carex sedge | *Carex* spp. |
| Oreobolus sedge | *Oreobolus pectinatus* |
| Rush | *Carpha alpina* |
| Rush | *Juncus* spp. |

*Forbs*

*Cushions or mats*

| | |
|---|---|
| Alpine daisy | *Celmisia argentea* |
| In the Figwort family | *Chionohebe* spp. |

*Leafy or rosettes*

| | |
|---|---|
| Succulent in the Bellflower family | *Lobelia roughii* |
| Succulent in the Bluegrass family | *Poa sclerophylla* |
| Succulent in the Gentian family | *Gentiana divisa* |
| New Zealand flax | *Phormium tenax* |
| In the Madder family | *Coprosma perpusilla* |
| Sundew | *Drosera* spp. |
| Bladderwort | *Utricularia monanthos* |
| Cushion speargrass | *Aciphylla dobsonii* |
| Fireweed | *Epilobium* spp. |

*Cryptogams*

| | |
|---|---|
| Cladina (Fruticose lichen) | *Cladina* spp. |
| Cladonia (Fruticose lichen) | *Cladonia* spp. |
| Sphagnum moss | *Sphagnum* spp. |

## Some Characteristic Animals of the Southern Alps, New Zealand

*Herbivores*

| | |
|---|---|
| Red deer | *Cervus elaphus*[a] |
| Chamois | *Rupicapra rupicapra*[a] |
| Tahr | *Hemitragus jemlabicus*[a] |
| Goat | *Capra hircus*[a] |
| Wapiti or Elk | *Cervus canadensis*[a] |
| European hare | *Lepus europaeus*[a] |

*Birds*

| | |
|---|---|
| Flightless Rail or Takahe | *Porphyrio mantelli* |
| Alpine Parrot or Kea | *Nestor notablis* |
| New Zealand Falcon | *Falco novaeseelandiae* |
| Australasian Harrier | *Circus approximans* |
| Rock Wren | *Xenicus gilviventris* |
| Pied Oyster Catcher | *Haematopus ostralegus finschi* |
| Banded Dotterel | *Charadrius bicinctus* |
| Southern Black-backed Gull | *Larus dominicanus* |

*Reptiles and amphibians*

| | |
|---|---|
| Skink | *Leiolopisma* spp. |
| Gecko | *Hoplodactylus* spp. |

*Note:* [a]Introduced.

# 4

# Tropical Alpine Biome

High-elevation tropical ecosystems, generally between $23^{1}/_{2}°$ N and $23^{1}/_{2}°$ S, are the least known of all tropical biomes. Geographically isolated, they are like cold islands surrounded by warm wet rainforests. Tropical alpine zones are distinct from their arctic and mid-latitude counterparts in both vegetation and climate, and the use of the term tundra is misleading because of numerous differences. Most areas have unique geographic names but are too small to be shown on a world map. The tropical alpine environment in the northern Andes, straddling the Equator from Venezuela to northern Peru with a few outliers in Costa Rica, is called the páramo, while farther south in the central Andes, the altiplano is also known as puna. The puna is the largest area of tropical alpine habitat in the world; it ranges from central and southern Peru and Bolivia to northern Chile and Argentina. In East Africa, it is called the afroalpine zone, and in Indonesia and New Guinea it is simply called tropical-alpine. High zones on Hawaiian volcanoes are also above tree growth. Unless a particular geographic region is specified, the term tropical alpine zone is used here to refer to all high-elevation ecosystems on tropical mountains. The lower limit of the zone is often difficult to define because there may not be a recognizable treeline. The alpine zone is above rainforest or cloud forest and treeline may consist of clumps of shrubs or stunted trees draped with epiphytes surrounded by grassland. On dry slopes where there is no forest, shrubs and succulents give way to typical alpine vegetation.

Although taxa on tropical mountains vary considerably from place to place because the high mountains are isolated from one another, similarities in climate, growthforms, and general environment exist. The páramo is similar to the afroalpine

**Island Biogeography**

Within the discipline of Geography, a subfield called Island Biogeography strives to explain the distribution of species on oceanic islands. Because of limited habitat and long isolation on oceanic islands, a unique biota often develops through the adaptive radiation of the few organisms that can make it far out to sea. The Galapagos and Hawaiian Islands are prime examples of places where plant and animal immigrants evolved to fill vacant niches. The same concept can be applied to island-like areas, such as high-elevation mountain top environments that have never been connected. They rely on similar types of long-distance dispersal and evolution and often have many endemic species.

in terms of plant growthforms. Like the isolated mountains in Africa and Indonesia, the páramo is not continuous but consists of smaller isolated patches in Venezuela, Colombia, Ecuador, and northern Peru. The puna has no Old World equivalent. In terms of growthforms, it is most similar to the mid-latitude alpine region of Tibet, but in terms of continuous extent, the puna is most similar to the Ethiopian Plateau. High mountain grasslands dominated by bunchgrasses with a scattering of other plants, especially shrubs and rosettes in the sunflower family, are typical of all tropical alpine areas. Some localities are purely grassland, while others have ground covered by giant rosettes and little grass. Locally important families with significant genera include pineapple (genus *Puya*) in South America, bellflower (genus *Lobelia*) in eastern Africa, sunflower (genus *Senecio*) also in eastern Africa, rose (genus *Polylepis*) in the Andes, and tree fern (genus *Cyathea*) in New Guinea. Frost polygons and stripes may be found at the highest elevations. Ground cover varies from barren with occasional mats or cushions to a spongy mat of mosses.

Tropical mountains are geologically young, building primarily since the late Cenozoic. Therefore most of the biota has only recently evolved, and most species were derived from lowlands as mountains uplifted. Of about 300 alpine plants on East African volcanoes, 80 percent are endemic, having evolved from surrounding lowland vegetation. Similar development has taken place in other tropical mountains such as the Andes.

## Physical Environment

### Climate

Except for wet and dry seasons, tropical mountains display daily changes rather than seasonal ones (see Figure 4.1). Because of their tropical locations, daylength and temperature vary little throughout the year. Day and night are each consistently 12 hours long. No seasonal change in sun angle like that experienced in the mid-latitudes occurs, but sun angle differs according to slope aspect and steepness. Because the sun crosses the Equator in its path between the Tropics of Cancer and Capricorn, either the northern or southern slope can receive direct rays while the other is in shadow, depending on time of year. No seasonal variation in temperature and no long period of cold and dormancy occur. However, diurnal temperature changes are pronounced. During the day, tropical alpine sites are bombarded with solar radiation, intensified by the thin atmosphere at high elevation. Much heat is lost at night as energy radiates back to space. The difference in temperature

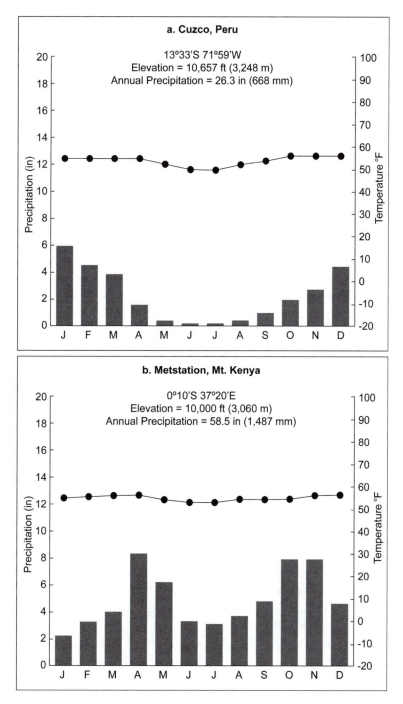

**Figure 4.1** Tropical alpine climates have similar temperatures but may vary in total and seasonality of precipitation. (a) Cuzco, Peru, is in the dry puna, but (b) closer to the Equator, the meteorological station on Mount Kenya receives more precipitation. *(Illustration by Jeff Dixon.)*

from day to night in tropical alpine areas is much greater than the difference between winter and summer. Consequently, plants must able to withstand temperature extremes every day of the year.

Mean annual temperature for all tropical alpine environments is about 50° F (10° C) at about 9,800 ft (3,000 m), the same temperature value that is often used to define the boundary between boreal forest and arctic tundra. The boundary also roughly coincides with the lowest elevation occurrence of freezing temperatures on wet slopes. Frequency of frost increases from a few days a year at 9,800 ft (3,000 m) to 100 days at 14,700 ft (4,500 m) and daily at the level of permanent snow above 15,400 ft (4,700 m). On dry mountains, frost begins slightly lower, but permanent snow is higher, at 19,700 ft (6,000 m), probably due to less precipitation. As temperatures decline with increasing latitude such as on volcanic mountains in Mexico at 19° N, the frost zone and treeline both begin lower. The coincidence of freezing temperatures with treeline in tropical mountains is significant for life. Plants must not only be adapted to temperatures that are consistently low, but also must be permanently frost resistant. Unlike plants in mid-latitude environments—alpine or not—that have different frost tolerances in summer than in fall and must winter-harden before they are able to withstand freezing conditions, tropical alpine plants may be subjected to freezing on any night of the year and must always be prepared. Plants must also photosynthesize under both low temperatures and low light conditions because days are frequently cloudy.

Tropical mountains experience rainy seasons and less rainy or even dry seasons lasting up to six months. In equatorial areas where precipitation is controlled by convection, rainy seasons occur twice a year during the equinoxes, when the sun and equatorial low pressure belt pass overhead. The dry seasons coincide with the solstices when high pressure dominates. Climates dependent on trade winds to bring moisture usually have only one rainy period related to seasonal shifts in the low pressure belt. Most precipitation is due to thunderstorms.

While temperature predictably decreases with elevation, changes in precipitation are more complex. In the Tropics, precipitation increases to a maximum at mid-level elevations where cloud forests occur, but then steadily decreases toward the mountain summits. Stable air above the cloud forests prevents storms from rising farther upslope. Therefore, alpine areas are considerably drier than lowlands in the same latitude. The zone of maximum precipitation is generally between 3,000 ft (900 m) and 4,600 ft (1,400 m), but varies according to geographic location. Maximum rainfall is reached at 2,500 ft (750 m) in Hawaii, 3,150 ft (950 m) in the tropical Andes, and up to 5,000 ft (1,500 m) in eastern Africa. After the maximum, precipitation decreases with elevation at a rate of 4 in per 330 ft (100 mm per 100 m). It is difficult to generalize about precipitation totals for tropical alpine zones because several variables such as slope aspect and position with regard to equatorial low pressure or trade winds create different patterns of precipitation and rainshadow. Such variations contribute to diversity and influence the distribution of environments and plant communities.

Tropical alpine zones may have seasonal temperature regimes that coincide with the rainy and dry seasons. Periods of rain and clouds depress diurnal temperature changes, because less solar radiation enters during the day but also because more infrared radiation (heat) is trapped at night. Under clear skies, both high and low temperatures are more extreme. Most frosts occur during the drier season because with less cloud cover more infrared radiation is lost at night, causing temperatures to drop. Minimum temperatures—especially those below freezing—and maximum temperatures are more important to plant life than are daily, monthly, or annual means.

Tropical mountains rarely experience the strong winds common to mid-latitude mountains under the jet stream and cyclonic storms. It may be windy on exposed ridges, but not everywhere.

## Soil

Tropical alpine soils are similar to those in other alpine regions: young and poorly developed, usually bearing chemical resemblance to parent material. Steep slopes tend to be covered by scree and talus, while more level areas may have an accumulation of sediment and organic matter from plants. Surface soils may be dry, but deeper moisture is usually available. Landscapes in the high Andes south of Ecuador appear very dry with only a thin plant cover, but plants extend roots more than 3.3 ft (1 m) into the soil and well-established plants have access to soil moisture. Coarse and dry cinders on volcanic mountains prevent evaporation from moist sand or soil beneath. Surfaces are subjected to temperature extremes on a daily basis, but temperatures at depth are more stable. Soil temperatures 3 in (8 cm) deep on Mount Kenya are constant at about 37° F (3° C). The surface, however, is subject to frost-heave. Unlike arctic tundra and mid-latitude alpine environments, tropical alpine areas have no permafrost, but they do experience small-scale solifluction where nightly freezing disrupts soil development. In bare soil, needle ice forms nightly in the afroalpine and other tropical alpine zones, disturbing the soil and making it difficult for seedlings to survive.

## Plant Adaptations

Tropical alpine areas support five growthforms, each dominating a different environment largely based on soil and moisture availability—tussock grasses, cushion plants, sclerophyllous shrubs, ground-level rosettes, and giant stalked rosettes (see Figure 4.2). Many growthforms are the same as in arctic tundra or in mid-latitude alpine regions, but others have evolved unique ways to endure the extreme temperature range.

Low-stature plants, such as stalk-less rosettes, tussock grasses, and cushion plants, create a mild microclimate within their foliage at ground level, thereby avoiding harsher air conditions. Grass tussocks have a layer of dead vegetation that

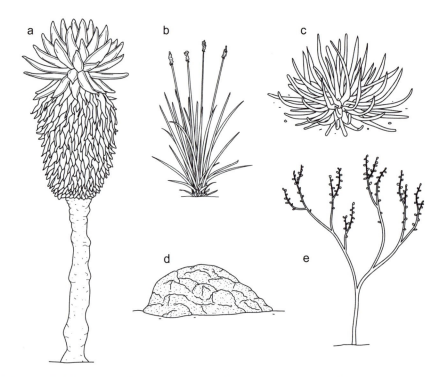

**Figure 4.2** Major growthforms in tropical alpine regions include (a) giant rosettes, (b) tussock grasses, (c) ground-level rosettes, (d) cushions, and (e) shrubs. *(Illustration by Jeff Dixon.)*

both insulates and holds water. After a fire, flowering and regrowth quickly follow to replenish this necessary insulation. The masses of stems in tussock grasses also support the plant against wind. Hairiness is a common characteristic and serves several purposes. In addition to shielding the plant from intense sun, hairs provide insulation at night, which helps retain heat. Air trapped by the hairs maintains its humidity, thereby reducing the transpiration loss during the hot, dry day. Tropical alpine plants, especially large ones, often have a silvery color that reflects excessive solar radiation. Shrubs with leathery or waxy (sclerophyllous) leaves are well adapted to periodic droughts and intense radiation in alpine zones. Small leaves present less surface area to the cold, and a waxy cuticle is able to withstand dry air conditions. Many high-elevation plants, such as giant lobelias in the afroalpine, have a red pigment that protects against strong radiation. The number of plants that are either hairy or have red coloration increases with elevation. Many plant adaptations, such as hairiness or waxy leaves, make then unpalatable to herbivores, providing an additional survival trait.

Little information is available about reproduction by seed among tropical alpine plants. Flowering takes place all year on Mount Kenya, but most occurs in January and July following the rainy seasons. In the Andean páramo, however,

**Figure 4.3** Although not related, giant rosettes have similar growthforms: (left) giant groundsel (*Dendrosenecio keniodendron*) in Equatorial Africa and (right) espeletia in the Andean páramo in El Angel Ecological Reserve, Ecuador. *(Left, courtesy of Rainer W. Bussmann, Missouri Botanical Garden, and right, Susan L. Woodward.)*

seasonal drought limits flowering time. Flies, bumblebees, and wind are important pollinators, although two species of hummingbirds are known to pollinate espeletias. Several plants require cross-pollination. Espeletia seeds need a cold period of 30 days at 35° F (2° C) before germination. Giant rosettes in the Tropics are long-lived plants. Silversword in Hawaii does not flower until it reaches age 20 and it lives to be 90 years old. One of the lobelias (*Lobelia telekii*) on Mount Kenya flowers and dies at 40–70 years, while the much larger giant groundsel may live for several hundred years.

## Giant Rosettes
Although not present in every habitat, a conspicuous growthform unique to tropical alpine tundra is giant rosettes. Giant rosettes include espeletias and puyas in the Andes; lobelias, giant senecios, and giant groundsels in Africa; tree fern in New Guinea; and silversword in Hawaii (see Figure 4.3). Giant rosettes, whether ground-hugging or stalked, have unique ways of coping with the tropical alpine climate. Stalks or inflorescences can be 3–30 ft (1–10 m) tall. Plants usually have a sheath of dead leaves, 4–12 in (10–30 cm) thick, which insulates the stem from cold, traps water, and provides humus. Temperature inside the stem remains above

the freezing point. The chemical composition of the leaves changes, indicating that nutrients are transferred from dead leaves to new ones, which is an efficient way to recycle scarce soil resources. In taller rosettes in East Africa, the outer leaves always enclose the central growing point. These leaves close at night. The closer temperatures are to freezing, the more tightly the leaves close; on warm cloudy nights, the rosettes only loosely enclose the growing point. Flower buds may be protected the same way but not only at night. Leaves may curl over the bud whenever weather turns cloudy and cool. One of the giants (*Senecio brassica*) in the upper alpine zone of Mount Kenya has an extra advantage. It supplements its insulation with a dense wooly covering on the underside of its leaves. When a rosette closes, this "wool" provides extra warmth. The flowers on tall inflorescences, which are exposed to air temperatures farther above ground level, are protected from extremes of temperature by insulating bracts or hairs, or they are sunken into the stalk. Buds on puyas in the Andean puna are protected deep within a hairy inflorescence.

A unique method of preventing freezing has developed among the giant lobelias on Mount Kenya. The broad leaf base catches water. The plant secretes a fluid that lowers the freezing point of the water, keeping it liquid. The lobelia requires a wet habitat so the prevention of freezing is necessary. A small fly takes advantage of the warmer site and lays its eggs in the water, thereby protecting its larvae from freezing.

Stems of both Andean espeletias and the African giant rosettes contain a pith that stores water for use during times of low rainfall or when soil water is unavailable because the ground is frozen. So much water can be held that the mountain gorilla of southwestern Uganda will snap off rosettes of giant groundsels to obtain the watery pith. Plants from wetter habitats or warmer lower elevations have less pith.

Giant rosettes also grow in tropical alpine zones where the surface soil may be quite dry. The plants must be able to pull water up and through tall stalks. Even though dead leaves keep the plant warmer than outside air temperatures, the xylem (tissue that transports water from roots to leaves) is susceptible to freezing at night. When night temperatures are above freezing, giant groundsels and tall lobelias on Mount Kenya do not alter transpiration the next day. After a hard frost, however, some do decrease transpiration, perhaps because the xylem is too cold or frozen, causing a temporary physiological drought. Many giant rosettes have short stems so the water has less chance of freezing overnight. Giant rosettes, however, rarely suffer moisture stress because deep roots tap deep soil moisture.

····························

**Nature or Nurture?**

Intense solar radiation or temperature changes may be the reason for rosette growthform in Tropical Alpine Biomes. Alpine plants were moved from 13,700 ft (4,175 m) on Mount Kenya to a lab in Nairobi at an elevation of 5,327 ft (1,624 m) and kept at 75° F (24° C). After only 10–27 days, leaves grew twice as large and stems elongated, destroying the compact rosette form. Plants subjected to cold night temperatures, however, did not change.

····························

## Treeline Flora

Treeline flora, as well as alpine flora, in the Tropics is diverse because mountains are isolated with little to no past or present connections to each other. Treeline in Old World tropical mountains is dominated by members of the heath family—evergreen shrubs with tough, leathery leaves able to regenerate after fires. Tree heather and *Philippia* heath species form shrublands in Equatorial East Africa. Shrubby *Polylepis*, in the rose family, grows at high elevations of the tropical Andes, and true trees can be found at 13,800 ft (4,200 m) or even higher. Evergreen alder ranges from Mexico to northern Argentina and forms treeline where *Polylepis* is absent.

It is difficult to define treeline in tropical alpine mountains because giant groundsels in eastern Africa and espeletias in South America can be called "trees" because of their erect woody stems. In some regions, such as the west slope of the Andes, desert conditions extend high into the mountains, and where there is no forest, there can be no treeline. The lower limit of the alpine zone is often better designated by climate conditions and plant growthforms.

## Animal Adaptations

Faunas of tropical alpine environments are usually dominated by only a few families or genera. Individuals are difficult to see in the field because they are most often small and have dull colors, unlike lowland animals in the Tropics. Most are also secretive, hiding in burrows or among rocks.

In spite of the decrease in both air density and oxygen, birds do not appear to have special adaptations to high-elevation environments. They are already physiologically adapted to maximize intake of oxygen in their breathing mechanisms. Lower air density does, however, mean that less lift is available and more energy needs to be expended to fly. Birds with great wing area or smaller body size are at an advantage. So in a seeming paradox, very small birds—such as New World hummingbirds and Old World sunbirds—and very large soaring birds—such as condors and Old World vultures—are the most common birds at high elevations. Hummingbirds have another advantage in that they can survive a lowered body temperature, called torpor, and achieve a corresponding decrease in metabolism. A bird will go into torpor only on those nights when energy and food supplies are low. Nectar-feeding sunbirds in the Old World and hummingbirds in the New World are common pollinators in their respective tropical alpine areas.

In the Andean páramo, temperature can fluctuate up to 45° F (25° C) each day, and nightly lows reach 20° F (−6.5° C). Tropical alpine insects must survive or avoid both nightly freezing and daily high temperatures with possible desiccation. Insects shelter in rock scree or in vegetation, especially in dead leaves of espeletia rosettes where temperatures remain above freezing. As many as 130,000 insects have been estimated to live on a 4 ft (1.2 m) stem, with different species occupying different parts of the plant. Most páramo insects cannot supercool and would freeze without shelter. However, some grasshoppers stay in open vegetation between rosettes and survive low ground night temperatures because they are tolerant of freezing. Similar patterns of insects occur in and around giant senecios on Mount Kenya.

## Regional Expressions of the Tropical Alpine Biome

### South America

Treeline in the Andes Mountains is about 11,000 ft (3,250 m), above which is the tropical alpine zone (see Figure 4.4). High-elevation landscapes in the northern Andes, 10° N to about 5° S, are called páramo. From northern Peru through Bolivia to northern Chile and Argentina, 5°–22° S, high inland basins are called puna or altiplano. Although both are basically grasslands, major differences include amount of rainfall, height of grasses, associated plants, and animal life. With up to 75 in (1,900 mm) of precipitation, páramo is high-elevation wet grassland, while puna with less than 27 in (685 mm) is high-elevation desert grassland. Giant

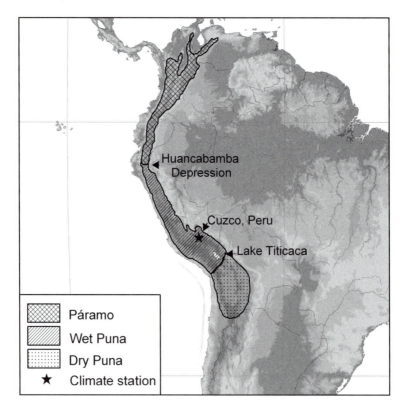

**Figure 4.4** Alpine environments in tropical South America. *(Map by Bernd Kuennecke.)*

espeletia rosettes characterize parts of the páramo but are absent from the puna. Large animals such as guanaco and vicuna graze the puna but not the páramo. Landform differences are evident as well. In contrast to rugged alpine glaciation and snow-capped volcanoes typical of the páramo in the northern Andes, the puna is characterized by broad intermontane basins or "altiplanos."

*Páramo.* The southern border of the páramo is the great Huancabamba Depression where the upper Marañón River cuts deeply through the Andes in northern Peru. While creating a corridor for species passing between the Amazon and the Pacific coast, the low elevation is a significant geographic barrier for passage of high Andean flora and fauna between north and south.

The páramo from Venezuela through Colombia to Ecuador is a moist grassland, an herbaceous ecosystem lying between forest at elevations below 9,850 ft (3,000 m) and permanent snowline above 15,000 ft (4,500 m). It generally has low mean temperatures with night and morning frosts and high humidity. The Spanish name páramo, referring to sterile and uninhabited lands, reflects its treeless nature. It is an area of steep mountains, with typical alpine glacial landforms. In Venezuela, much of the páramo occurs as isolated islands above cloud forest. The ice layer in the soil is not deep like arctic permafrost, and melts daily, saturating the soil and causing solifluction. No seasonal changes occur, so the growing season lasts all year. The páramo is dominated by grasses and rosette plants. With its tall bunchgrasses or tussock grasses and giant rosettes, the Andean páramo is similar to the alpine zones of Mount Kenya, especially in the resemblance in appearance of some of the giant rosette plants.

The páramo climate is generally moist, with almost daily presence of mists. Precipitation at 9,850 ft (3,000 m) varies from 27 in (685 mm) to 75 in (1,900 mm) depending on location with respect to the Equator and position on the windward or lee side of the Andes. It is only 36 in (900 mm), however, at higher elevations. There may be either one or two rainfall seasons, but the driest months are January and February north of the Equator and June and July south of the Equator. The rainy seasons, April through June and August through November, respectively, coincide with times of the year when the equatorial low pressure is near the Equator. Low precipitation in the dry season, however, makes parts of the páramo like deserts for four or more months due to both evaporation and runoff on steep mountains.

Temperature on the páramo can change dramatically throughout the day. At night, temperatures drop below freezing, but they can rise to 73° F (23° C) by midday. Many days, however, have mean temperatures around freezing. Glaciers still cover several peaks, but they have been retreating rapidly in the last decade. With little difference in daylength throughout the year because of its equatorial location, the amount of sunlight received does not vary from season to season. Clear skies versus cloud cover, however, affect the amount of incoming radiation reaching the surface during the dry season and in the morning before clouds form. Upslope winds cause cloud formation by early afternoon, blocking the intense equatorial sun and chilling the air. Plants must be ready to withstand variations in solar radiation, extreme heat, and freezing temperatures on a daily basis.

Temperature varies according to height above the ground and cloud cover. During the dry season, temperatures 5 ft (1.5 m) above the ground average 27° F (−2.8° C) at night compared with 52° F (11° C) during the day, and temperatures remain below freezing for 13 hours, from 8 PM to 9 AM. Temperature variation even 4 in (10 cm) above the ground is similar, but variation in soil surface temperatures is extreme, ranging from 14° F (−10° C) at night to sometimes as high as 104° F (40° C) during the day. During the wet season when it is cloudy, all temperature variations are less extreme.

Soils are shallow with few nutrients and low water-holding capacity. Above 13,100 ft (4,000 m), in what is known as the superpáramo, a daily freeze-thaw cycle leads to solifluction, resulting in unstable soils. At slightly lower elevations, nightly frosts create needle ice. Most soils are poorly developed and are well-drained entisols or inceptisols. However, some areas are inundated with water all year, resulting in bogs with sphagnum moss. Most soils are acidic.

*Vegetation.* Even though by definition the páramo is treeless, high Andean woodland groves do occur. *Polylepis* trees grow in small stands above treeline, forming vegetation islands where protected microclimates are warmer. In these areas, wet scrublands consisting of reedgrass, bentgrass, St. John's wort, as well as *Polylepis* trees develop.

The flora is rich. Some 420 kinds of flowering plants occur in the Venezuelan páramo, and more than 230 are in the superpáramo. With no direct connection now or ever with the Arctic, few arctic plants have reached the Andes, and most species, some of which are unique to a particular peak or cluster of peaks, derive from local lowland floras. A major exception is Iceland purslane, which dispersed from the Arctic south via the Rockies and Andes all the way to Tierra del Fuego. This plant may have had a dispersal advantage because it is an annual and its seeds can be transported long distances by wind or migrating birds. Some forbs, such as geranium, gentian, paintbrush, buttercup, and others, are represented at the genus level in both the Arctic and mid-latitudes.

The páramo contains several plant communities. The most common growthform is bunchgrass. The vegetation has two layers, an upperstory of bunchgrasses and an understory of mats, cushions, rosettes, creepers, subshrubs,

· · · · · · · · · · · · · · · · · · · · · · · · · · · · · · · · ·

### *Frailejones*

Several rosettes in the sunflower family, especially espeletias—called *frailejones* in Spanish, meaning gray friars—are prominent in the páramo and superpáramo of Venezuela, Colombia, and northern Ecuador. Some are tree size, up to 10 ft (3 m) tall, while others are only a few inches high. Most espeletias, grow at elevations between 9,800 and 13,100 ft (3,000–4,000 m), but some may be found as low as 7,200 ft (2,200 m). Espeletia has a tall stem or trunk with leaves at the top. Old leaves die and drape down over the stalk. Plants depend on insects, birds, and wind for pollination.

· · · · · · · · · · · · · · · · · · · · · · · · · · · · · · · · ·

**Figure 4.5** The páramo in the northern Andes is predominantly tussock grasslands. *(Photo by author.)*

lichens, and mosses. Grasses are diverse, with 30 species in the Venezuelan páramo. Most are bentgrasses, bromes, reedgrasses, fescues, and needlegrasses. Mosses, liverworts, and lichens are abundant and diverse. In spite of the moist climate, the páramo always looks as though it were autumn because dry vegetation among new growth lends a brownish tinge to the landscape (see Figure 4.5). Wet grasslands with tall bunchgrasses or tussock grasses often include giant rosette espeletias and puyas. Boggy or swampy areas support large cushion plants, while forbs, including alpine orchids, occupy open rocky areas. Dwarfed bamboo, shrubs from several families (sunflower, mahogany, and St. John's wort), and sedges also occupy these areas. The moist ground may be covered with a dense mat of mosses, lichens, and cushion plants. Wet, marshy soils, with rocky outcrops of volcanics and other rock types, are the headwaters of streams. Notable flora in marshy areas includes quinine shrubs.

Above 12,800 ft (3,900 m), in the desert páramo, short cushion plants become more prominent. Around 14,000 ft (4,275 m), grasses decrease in abundance and low shrubs, mats, and rosettes begin to dominate. Sandwort cushions form dense colonies only a few inches high. In some places, large cushions of endemic *Azorella crenata* and *A. julianii* grow with tall espeletia rosettes. With age, azorella cushions can grow to 3.3 ft (1 m) high. The Andean desert páramo is primarily found in the Cordillera de Mérida in Venezuela, from 13,100 ft (4,000 m) up to the edge of glaciers at 15,100 ft (4,600 m). Extreme habitats at this elevation are called periglacial, characterized by daily freeze-thaw cycles, small-scale patterned ground, and solifluction. These highest desert periglacial zones have sparse vegetation.

Three plant associations with giant espeletia rosettes occupy periglacial habitats, each with a different species. One, up to 10 ft (3 m) tall and topped by an inflorescence of 40 in (100 cm), is most widely distributed above 13,000 ft (4,000 m), on steep slopes with unstable fine scree. Colonization stabilizes the slope, and plants may live for 170 years. A different, smaller espeletia grows on rocky sites such as aretes and cirque walls where large boulders retain heat, allowing the plants to survive at higher elevations. This species grows only 3 ft (1 m) tall and has a short inflorescence; it also has the shortest life span of the three, 70 years. A medium-size espeletia grows 7 ft (2 m) tall on gentle slopes with medium-size gravels. Its inflorescence is 30 in (75 cm) long and the plant lives for 130 years. Long life is an advantage because these three species only reproduce by seeds. Each plant produces 1 million seeds in its lifetime. Two other espeletias have prostrate stems forming giant cushions or shorter rosettes, respectively. These two are more important in the lower Andean zones but grow up to 13,800 ft (4,200 m) in wetter parts of the desert páramo.

The three high-elevation páramo espeletias continue to grow after flowering. Their inflorescences sprout from between young leaves near the top of the rosette, even as the terminal bud continues to grow upward. Espeletias that die after flowering because the terminal bud becomes the flower stalk are found only below 13,000 ft (4,000 m).

Many plants at elevations of 14,000 ft (4,275 m) or higher have dense, wooly hairs. One type of espeletia, a large rosette of thick, wooly, white leaves on a stalk draped with the old dead leaves, grows at this elevation. This plant, the largest espeletia species, can grow up to 15 ft (4.5 m) tall and is similar in appearance to the giant groundsels of Africa. It grows tallest in wet areas, near streams, lakes, and peat bogs and makes a kind of savanna with the grasses.

Other habitats in the high alpine zone include loose sand and gravel. Two shrubs, chuquiragua (or flower of the Andes) and jata (or candlebush) are pioneers on this substrate, growing until sufficient soil has developed for grasses to take root. Occasional rock scree in the upper alpine within the permanent rock and snow at 13,500–15,750 ft (4,100–4,800 m) supports only scattered small rosette plants.

*Animals of the páramo.* The few mammals inhabiting the páramo are primarily small herbivores, including páramo rabbits and several rodents. Rabbits are found mostly under cover of St. John's wort. Their burrows are in two sections, a small one for refuge or escape, and a larger one used as the nesting chamber. Nests are constructed of dry grass, moss, and hair, which insulates against cold temperatures. Adaptations to the páramo include heavy fur, long gestation period, small litters, and few teats. Grasses, especially bromes, are the rabbits' main food. Rabbits are preyed on by Black-chested Buzzard Eagle, White-rumped Hawk, owls, and an occasional puma and páramo wildcat as well as by feral dogs.

Larger mammals of note include mountain tapir, little red brocket deer, northern pudu, and the omnivorous spectacled bear. The mountain tapir is the only tapir that does not live in tropical forests. Little red brocket deer and northern pudu are both small ungulates that usually live in montane forests but, because of their small size, can also take shelter in grassland. Brocket deer measures 28 in (70 cm) tall at the shoulder, and pudu, the world's smallest deer, is only 14 in (35 cm) high. White-tailed deer are also found all along the Andes chain, living on valley floors in small family groups of up to five individuals. Although they prefer *Polylepis* stands, they also inhabit open and shrubby páramo. The only other large herbivores are domesticated llamas and alpacas.

Thomas' small-eared shrew is significant in the páramo ecosystem, building a complex system of nests and tunnels beneath espeletias and St. John's wort. It preys on young rodents, eggs and chicks of ground-nesting birds, lizards, and insects, and is itself prey for opossum, weasel, eagles, hawks, and owls. Long-tailed weasel is another major small predator. Puma and páramo wildcat occasionally may be seen.

The most commonly seen animals are small birds. The high elevation of the páramo makes it a convenient resting stop for migratory birds. Among the breeding birds, nectar-seeking hummingbirds are important pollinators and have the advantage of inducing torpor when temperatures drop. Alpine birds generally build larger nests than do their lowland counterparts, and also choose sites that are more stable in temperature. With few predatory snakes and lizards and no trees (except scattered *Polylepis* groves) in the páramo, nests are built on the ground or in low bushes. Small insect-eating birds are most numerous. Ground-feeding insectivores include Páramo Pipit and Streak-backed Canastero, while Mérida Wren and Sedge Wren glean insects from small bushes. Fruit eaters, large seed eaters, and scavengers are not common because fleshy fruits, large seeds, and large carcasses are rare. Eagles are present but scarce, as are Andean Condors. Condors, extinct since the 1950s in Venezuela, are being reintroduced, but there is concern that too few large animals are available for food.

Many insects, birds, and even a small rice rat live in the sheath of dead leaves that cover giant rosette stems. Simultaneous September to December flowering of most giant rosettes attracts many pollinators. More then 1,000 insects were collected around espeletia flowers in the Venezuelan páramo. Most abundant were flies and midges, but bees and wasps were also well represented. Butterflies are conspicuous, and beetles are important both in numbers and species diversity, with about 25 families represented in the Andean highlands. Carabid beetles, in particular, have many species in the páramo.

Although espeletia has chemical toxins to deter grazing by native animals, nonnative grazers are not affected. Much of the páramo has been converted to agricultural land and is subject to overgrazing, a significant threat to the ecosystem.

*Puna.* From the Huancabamba Depression south to western Bolivia and northern Chile and Argentina is the puna, at a general elevation of 9,850–16,400 ft (3,000–5,000 m). The high ranges to the east face the easterly trade winds, and the puna in their rainshadow is dry. Puna is cold grassland with drought-adapted bunch grasses. Needlegrass is prominent, along with shrubs of the sunflower family. Other common plants are fescue and reedgrass, with prostrate rosettes. The eastern puna is more moist and has a closed grassy vegetation, while the central and western parts are dry and shrubby. Precipitation and plant growth patterns reflect the occurrence of a single rainy season and an intense winter dry season. The wetter puna in the east receives 30 in (750 mm) of rainfall. Conditions become drier both to the west and south, until extreme drought is reached in the Atacama Desert at the border of Bolivia, Chile, and Argentina. The decrease in precipitation is especially noticeable south of 15° S. Seasonality also increases toward the south, in the form of a longer dry season. About 85% of annual precipitation falls in the rainy summer, with almost none during the dry winter. Temperatures also vary according to latitude. Toward the south, increasingly greater seasonality and colder winter temperatures are experienced. Temperatures of 5° F (−15° C) have been recorded in the altiplano in southern Peru and −22° F (−30° C) in the Bolivian puna. The southern limit of the puna at 20° S lies where the annual temperature range increases to 18° F (10° C), and winters are more severe. The altiplano has been affected by livestock grazing for centuries, and it is questionable as to whether the current vegetation represents natural landscape or managed pasture.

The wet puna, located between 12,100 and 13,800 ft (3,700 and 4,200 m) from northern Peru to northern Bolivia, is covered with tussock grassland and shrubs (see Figure 4.6). The Andean mountains in this area divide into two ranges, the Cordillera Occidental and Cordillera Oriental. A central large plateau called the Altiplano lies between them. The high peaks and valleys above the Altiplano are covered with ice; glacial tongues may advance as low as 9,800 ft (3,000 m). Bentgrass, reedgrass, fescue, and needlegrass (ichu grass) are the most conspicuous grasses, with mountain bamboo and pampas grass in more humid areas. Poorly drained regions in this zone support sedges, rushes, and bulrushes.

The high Andean puna, above 13,800 ft (4,200 m) on the Altiplano, has extreme diurnal temperature changes because of the high elevation. Temperatures drop below freezing nightly. Annual precipitation frequently falls as snow or hail instead of rain. Grasses are fescue, Peruvian feathergrass, and reedgrass. Tussocks can be 3 ft (1 m) in diameter and just as tall. Prostrate or rosette plants include gosmore and azorella cushions (see Plate XIII). Cushion bogs called "bofedales" are wet areas above 13,000 ft (4,000 m), with large cushion plants submerged or floating in water.

The dry puna occurs in western Bolivia and northern Chile and Argentina. This is a vegetation of primarily tropical alpine herbs and dwarf shrubs rather than grasses. Characterized by low precipitation, 2–16 in (50–400 mm) yearly, the dry season here lasts eight months. This shrubland steppe has drought-resistant shrubs such as *Adesmia,* broom rape, and tolillar, or grassy steppe with reedgrass, fescue,

**Figure 4.6** The Altiplano, seen here in southern Peru, is a high-elevation plain between high mountains. Ampato Volcano is in the background. *(Copyright © Dr. James S. Kus.)*

and needlegrass. Shrubs may be 8 ft (2.5 m) tall. A vast inland sea on the Altiplano in this region left behind extensive salt flats, called *salares,* including Coipasa, Uyuni, Chalviri, and Arizaro (see Figure 4.7). Halophytic plants more commonly found in desert areas such as saltbush, saltgrass, pickleweed, and seepweed grow in this environment. Salt concentration in the tissue of saltbush is higher than that in the soil, which allows water to move from the salty soil into the plant. To prevent

······································································································

**Puya**

Although the genus is the only member of the pineapple family to grow in the high Andes, several species of puya are common giant rosettes in both the páramo and puna (see Plate XIV). Two of the most common are *Puya hamata*, which is an indicator species of the high páramo, and *Puya clava-herculis*. Rosettes of both species are generally small, less than 2 ft (0.6 m) in diameter, but the inflorescence can be 5 ft (1.5 m) tall. In contrast, the endangered *Puya raimondii*, which only grows at 13,100 ft (4,000 m) or above, can reach 10 ft (3 m) in diameter. After 80–100 years, this rare plant flowers once, producing an inflorescence 15 ft (4.5 m) tall before dying. Up to 20,000 small flowers on the stalk are protected from freezing by hairy bracts. Except for the giant *P. raimondii*, puyas resemble in size and form the lobelias of the afroalpine zone in East Africa.

······································································································

### San Pedro Cactus

Growing in the mountains of Ecuador and Peru at 5,000–9,000 ft (1,500–2,700 m), the San Pedro cactus can withstand cold conditions. It is a multistemmed columnar giant up to 20 ft (6 m) tall, with arms that can spread 6 ft (2 m) wide. Because its smooth green flesh contains small quantities of mescaline and other hallucinogenic compounds, it has a long history of ritual use in pre-Columbian times. Peruvian shamans, or "curanderos," still drink a mixture made from the plant. They believe it helps them diagnose a spiritual or subconscious basis for illness. The plant is legal to grow in most countries as an ornamental, but, with few exceptions, is illegal to use.

even more salt buildup inside the plant as evaporation takes place, the plant exudes salt onto the surface of its leaves. The salty surface gives the plant a grayish look and also renders it inedible to most herbivores. Old Man of the Andes cactus, named for its long white hairs, grows on islands in the Uyuni salt flats. Bofedales with large floating cushion plants also occur.

*Animals of the puna.* The most representative native mammals are vicuna, guanaco, chinchilla, and vizcacha. Members of the camel family such as guanaco, vicuna, and the domesticated relatives alpaca and llama have strong incisors that enable them to crop plants with high silica content (see Figure 4.8). Guanacos, probable ancestors to domesticated llamas and alpacas, are wide-ranging, but vicunas are restricted to high elevations above

**Figure 4.7** The vizcacha rat is adapted to feed on salty plants around salt flats in the Altiplano. *(Courtesy of Susan L. Woodward.)*

11,500 ft (3,500 m). The golden vizcacha is a rodent well adapted to browsing on saltbush. Like other rodents, it has both upper and lower incisors. It also has an extra upper pair of "teeth" that are actually made of stiffened, bristly hair, which allows the animal to scrape salt off saltbush leaves before consuming them. The animal's kidneys are specialized to excrete excess salt.

The Andean hairy armadillo, 12 in (30 cm) long with a tail half again its size, primarily feeds on small vertebrates and insects, but it will also eat roots or tubers. Hairs between its plates help keep it warm. Puma and Andean fox, are typical predators.

Darwin's Rhea, an ostrich-like bird standing 3 ft (1 m) high, is endemic to South America and found on the puna as well as on lower-elevation grasslands. Its diet consists of plants, seeds, roots, insects, and small vertebrates. Except during breeding season, the birds live in groups of up to 30 individuals. The male becomes very territorial when he incubates the eggs and cares for the young,

**Reed Boats**

Bolivian Indians have traditionally built boats from reeds growing around Lake Titicaca, although they are now replacing them with longer-lasting wooden boats. When Thor Heyerdahl first attempted to cross the Atlantic Ocean in a reed boat, Ra I, it failed just short of his goal in the Caribbean. A year later, in 1970, with an improved design and assistance from four Aymara Indians in Bolivia, Ra II was built. The journey from Morocco to Barbados was successful, covering 4,000 miles (6,450 km) in 57 days. The boat remained intact and seaworthy to the end.

**Figure 4.8** Vicuna, in Salinas-Aguado Blanca Reserve north of Arequipa, Peru, is frequently seen grazing the puna. *(Copyright © Dr. James S. Kus.)*

## Chinchillas

Due to demand from the fur industry for their soft, furry coats, these rodents were hunted almost to extinction, until the practice became illegal in 1929. Captive-raised animals that produce superior pelts minimized pressure on wild populations, but decimation of wild animals was so severe that chinchillas are still considered endangered or threatened. It is unknown how many remain in the wild. The Andean Mountain cat, the size of a domestic cat, is equally rare. Although its habits are largely unknown, it prefers high rocky mountains where chinchillas take shelter under rocks. It is likely that the cat may have depended on chinchilla for food.

## Flamingos

Three of the world's six flamingo species are found in the high Andes. In spite of frigid winter temperatures, hot springs that moderate water temperature help make the Altiplano salt lakes prime flamingo habitat. The attraction is an abundance of algae, brine shrimp, and other aquatic invertebrates. The rare and endangered James and Andean Flamingos, which live year-round in the Altiplano, are joined by the more abundant Chilean Flamingo during breeding season. With their beaks almost constantly moving through salt water while feeding, flamingos also ingest salt, which is excreted by special glands in their nostrils. The birds obtain fresh water from springs, puddles, or from their own feathers saturated with rain water.

often adopting orphaned chicks into his flock. Flightless, rheas are fast runners, attaining speeds up to 37 mph (60 km/h). Rheas escape predation by running a zigzag path or suddenly squatting in brush and flattening their bodies against the ground. The Puna Tinamou is another common ground bird. Many endemic birds, including Ash-breasted Tit-tyrant, Royal Cinclodes, three species of canastero, Olivaceaous Thornbill, and Gray-bellied Flower-piercer are widespread. Other endemic bird species are more locally distributed. *Polylepis* forests and scrub provide habitat for most of the endemic birds, found in both wet and dry puna. Large flocks of flamingo are found around the salt lakes.

## Afroalpine

Equatorial East Africa refers to Kenya, Uganda, Tanzania, and adjacent countries, while the term East Africa also includes Ethiopia, which has some dissimilarity in climate and landscape (see Figure 4.9). Afroalpine applies to all mountains within 10° to 15° N and S of the Equator that rise above treeline. With two exceptions, the Rwenzori Range and the Ethiopian Plateau, the afroalpine region consists of geologically young, isolated volcanic mountains that arose along the Eastern African Rift zone from Lake Malawi in the south to the Red Sea in the north. The largest volcanoes are the Virunga group along the western rift, and Aberdare, Mount Kenya, Mount Kilimanjaro, and Mount Meru on the eastern rift. Age varies from more than 15 million years for Mount Elgon and 2 million years for Mount Kenya to less than 200,000 years for Mount Meru, a factor affecting colonization and current vegetation. The Rwenzori Mountains consist of older, Precambrian intrusive igneous and metamorphic rocks. The Ethiopian highlands are a lava plateau broken by the East African Rift zone. At least 10 mountains exceed 13,000 ft (4,000 m). Mount Kenya at 17,058 ft (5,199 m) and Mount Kilimanjaro at 19,340 ft (5,895 m) are the highest. Glaciation has sculpted most peaks, and glaciers still exist today on the Rwenzoris, Mount Kenya, and Mount Kilimanjaro.

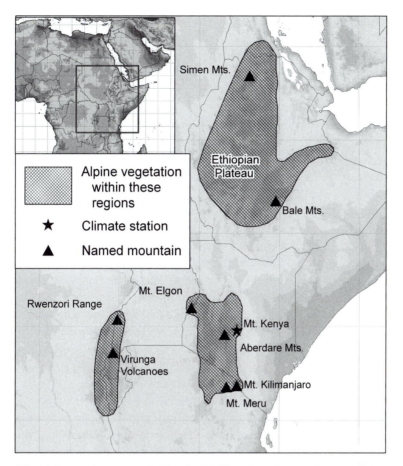

**Figure 4.9** Alpine environments in Tropical Africa. *(Map by Bernd Kuennecke.)*

The afroalpine flora is much less diverse than that of the Andes. Because high peaks are isolated and not a continuous chain, unique species occur on each mountain. More than 80% of the flora is endemic to the high mountains of East Africa, but not all species are confined to the afroalpine. With no corridors from the Arctic or from Central Asian highlands, few northern plants other than saxifrages and primroses are found. The flora instead has affinities with several other floral regions, including South African, Northern Hemisphere temperate, Mediterranean, and Himalayan.

***Climate.*** Few records exist, but differences in vegetation and zonation imply differences in climate. Two rainy and two dry seasons are typical. Winter—December through March in the Northern Hemisphere and June through August in the Southern Hemisphere—is driest. The majority of precipitation falls in the summer, April through May and November through December, respectively. Precipitation varies with exposure, and because moisture-laden winds come from the south and southeast, those sides of all peaks and mountain chains receive more rainfall. Total rainfall and lengths of wet and dry seasons vary considerably. The Rwenzoris are always moist, Mount Kenya has one month of drier conditions, and Mount Kilimanjaro has 4–11 months of drought. Annual rainfall in the alpine zone of Mount Kenya, more than 12,000 ft (3,650 m), is 35 in (900 mm), with less rainfall received at higher elevations. Mount Kilimanjaro is drier, receiving only 30 in (750 mm) at 12,500 ft (3,800 m), and only 5 in (125 mm) above 13,000 ft (4,000 m). The upper alpine zone of Mount Kilimanjaro is more barren than Mount Kenya because no sheltered valleys occur and the more porous substrate holds little water.

Although mean annual temperature is similar to that of both arctic tundra and mid-latitude alpine, major differences in the temperature regime exist. Major contrasts include a year-long growing season and a daily freeze-thaw cycle. Because of its equatorial location, the growing season is not limited by cold winters. As is true for other tropical environments, afroalpine zones are characterized by small seasonal changes and extreme daily temperature changes. On Mount Kenya at both 10,000 ft (3,000 m) and 17,000 ft (5,200 m), mean monthly temperatures vary only 3° F (1.5° C) throughout the year. Temperature decreases with elevation, from an annual mean of 45° F (7° C) in the forest at 10,000 ft (3,000 m) to 35° F (1.5° C) in tussock grassland at 13,750 ft (4,200 m), to 18° F (−7.5° C) in rocky areas at 15,650 ft (4,750 m) with glaciers nearby.

Due to both high elevation and equatorial location, intense solar radiation received during the day is lost as infrared radiation at night and temperatures drop below freezing. Rain and cloudiness affect temperature. Mount Rwenzori, with more clouds and rain, experiences only an 11° F (6° C) daily temperature range, while on Mount Kenya, where cloudy conditions are less common, the day-to-night temperature change is 22° to 32° F (12° to 18° C). On the more continental Sanetti Plateau in the Bale Mountains of Ethiopia, diurnal temperature changes are even more extreme. All these measurements, however, indicate the air temperature, not the microclimate at ground level. When air temperature at 14,800 ft (4,500 m) is 38° F (3.5° C), temperature at ground level or inside vegetation clumps can be a much warmer 67° F (19.4° C). Extreme differences between ground and air temperatures can be 56° F (31° C) at midday. Ground temperatures change rapidly in response to sunny or cloudy conditions. Under a cloud cover or in shaded sites, air and ground temperatures are similar. The lowest ground temperature measured in the lower alpine on Mount Kenya was 16° F (−9° C). Although such low temperatures are infrequent, heavy frost may occur nightly. Topography contributes to microclimate conditions, especially cold air drainage onto valley floors where more extensive solifluction disrupts soils and prevents colonization by plants.

Because of daily fluctuation in temperature, the relative humidity changes are extreme. Relative humidity of the air on Mount Kenya drops from 90% in the early morning to 20% or less at midday. At ground level, however, it remains fairly constant, at about 70%. Because of increasing relative humidity at night, moisture from dew and frost is significant. Wind has a local effect, and plants tend to grow in the lee of boulders near ridgetops and glaciers.

***Soils.*** Soils are derived almost solely from volcanic material but vary according to rock chemistry, ash deposits, slope, and precipitation. All are subject to serious erosion, often accelerated by overgrazing and agriculture. Fresh moraines are common parent material, coarse at valley sides and fine on valley floors where they retain water, leading to the formation of bogs. Because it is too cold for much chemical activity, soils are acid. Bare, steep slopes on valley sides are subject to solifluction. Small sorted polygons 5 in (13 cm) diameter on ridgetops above 14,000 ft (4,250 m) are the result of either frost action due to daily temperature fluctuations or to desiccation cracking. Finer material concentrates in the center while the edges are outlined by gravel. Mosses first take hold in the gravel edges, then are succeeded by grasses. Below 14,000 ft (4,250 m) where more tussock vegetation covers and insulates the soil, frost action is less common. Finer sediment on flat valley floors over 12,000 ft (3,650 m), however, is disturbed by needle ice, especially in bare soil between grass tussocks or on dried lake beds, preventing the establishment of plants.

***Vegetation.*** Altitudinal belts of montane forest, heath, and afroalpine vegetation are largely intact in Equatorial Africa; however, human-induced fires on Mount Elgon and Mount Kenya, have modified the afroalpine environment. Alpine zones in Ethiopia are highly disturbed by fire, wood collection, grazing, and agriculture. Montane forest, depending on location, primarily is composed of hardwood trees with a few conifers and often a bamboo zone. The heath vegetation is a distinct treeline zone, with trees and shrubs of tree heather and two *Philippia* species.

The alpine zone extends from the upper edge of the heath zone, about 12,000 ft (3,650 m) to the nival zone at 15,000 ft (4,500 m). Although no clear demarcation exists between the forest and heath belts and the alpine zone, few plants are able to survive the harsh afroalpine environment, resulting in a flora that differs from adjacent lowlands. Heath shrubs diminish in number and are replaced by giant rosettes and tussock grasslands as elevation increases. The upper limit of vascular plants, marked by an occasional strawflower, is 16,000 ft (4,875 m), but lichens are found on summits above 17,000 ft (5,200 m).

Three growth layers define afroalpine vegetation. An aerial level contains the giant rosette tree senecios, below which is an herb or shrub layer, usually of strawflower and grass tussocks. The ground layer consists of low rosettes and prostrate plants. Although the alpine zone is characterized by tussock grasses, the most obvious plants are lobelias and senecios—both giant rosettes—and lady's mantle and strawflower (see Plate XV). Many interesting alpine species are endemic to Mount Kenya or to East African tropical mountains in general.

All lobelias are low, compact rosettes with inflorescences up to 6 ft (2 m) tall. Woody, creeping stems remain prostrate along the ground. While *Lobelia keniensis* rosettes, endemic to Mount Kenya, grow from a single prostrate stem in the alpine zone, related species in the montane forest below have several erect branches. The growing points of alpine lobelias are suppressed by cold, allowing only one stem to grow. The rosettes close their leaves into tight buds at night to protect the growing point from cold. The stems and inflorescences of alpine lobelias are hollow, providing yet more insulation. While air temperatures drop to 23° F (−5° C), the core of the plant remains above freezing at 35° F (1.7° C).

*L. keniensis* commonly grows in groups connected by shallow underground stems. Leaves are broad without hairs, and the inflorescence grows up to 5 ft (1.5 m). The plants grow in moist areas and reach maximum density in the middle alpine elevations. A different lobelia, *Lobelia telekii*, endemic to Mount Kenya, Aberdare, and Mount Elgon, grows on well-drained, coarse substrate, primarily in the high alpine. It is differentiated by its narrow leaves with a waxy covering. Hairy bracts hide the flowers on the 6 ft (1.8 m) inflorescence.

Giant groundsels (genus *Dendrosenecio*) and giant senecios (genus *Senecio*) vary in growthform according to elevation, aspect, and microclimate, and dominant species vary with habitat. All continue to grow after flowering and all possess insulating layers of dead leaves around their stems. In several instances, people have used these leaf girdles as a ready source of fuel, killing the stripped plants. Giant groundsels are treelike with woody stems. Not all giant rosettes are upright. *Senecio brassica*, a giant senecio endemic to Mount Kenya and growing at lower elevations in the alpine zone than any other large senecio, has a low rosette of leaves, densely hairy on the underside, topping the end of a woody stem that creeps along the ground surface. The tall inflorescence, up to 3 ft (1 m), has bright yellow flowers. Some senecio species also occur as low, creeping herbs that can form mats up to 10 in (25 cm) high.

Of the several plant communities found in the afroalpine zone, the five most important are unevenly distributed. Although defined as above treeline, the afroalpine has distinctive woodlands or forests of tree-like giant groundsels, usually growing on deep soil with access to underground water. These giant rosettes occur on all high mountains except Mount Meru, where it may be too dry, and are one of the plants found at the highest elevations. The genus (*Dendrosenecio*) is endemic to Equatorial East Africa and absent from Ethiopia. Different but related species have evolved on each of the isolated mountains. Strawflower scrub also has different species in different locations. Found on rocky ground in all mountains, including Ethiopia, it is most impressive on the Rwenzoris, where it forms a dense scrub up to 6.5 ft (2 m) tall. Lady's mantle scrub is found on gently sloping, well-drained ground, and different mountains have different dominant species (see Figure 4.10). Dull brown tussock grassland dominated by fescue, but also containing bentgrass, bluestem, hairgrass, and *Pentaschistis,* is found on all mountains, but it is less common in wetter areas such as on the Virunga volcanoes and in the Rwenzoris. Grassland grows on well-drained, somewhat steep soils and replaces lady's mantle scrub

**Figure 4.10** Lady's mantle is a common shrub in East African alpine environments. *(Courtesy of Rainer W. Bussmann, Missouri Botanical Garden.)*

where the latter has been burned. Carex bogs form on flat or gentle slopes with poor drainage and usually create peat. Although found in the afroalpine, bogs are more important in the heath zone and are absent from Mount Meru. On wetter mountains, epiphytic mosses, lichens, and liverworts grow on the trunks of giant groundsels, especially where the leaf shield has been destroyed. Lichens predominate on drier mountains.

**Case Study: Altitudinal Zonation of Mount Kenya**

Mount Kenya is an extinct volcano on the Equator, with many other volcanic cones dotting its north and northeast slopes. Deep valleys, glaciated and U-shaped, radiate from its peak. Teleki Valley is one of the largest with smooth walls and craggy aretes. Steep valley walls, however, are not subject to erosion because surface scree allows water to drain through. Moraines are common. Mount Kenya's craggy summit and glaciated landscape with steep valleys contrasts with the smooth and rounded features of Mount Kilimanjaro.

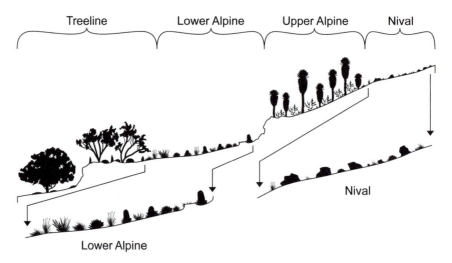

**Figure 4.11** Vegetational zonation on Mount Kenya. *(Illustration by Jeff Dixon.)*

The heath zone forms a less distinctive belt on Mount Kenya than it does on other East African mountains (see Figure 4.11), in part because it has been burned by human activities and only stunted shrubs remain. Composition varies due to different climate on different aspects. *Protea* may form closed communities up to 13 ft (4 m) tall, with undergrowth according to habitat. Tree heather and *Philippia* species up to 12 ft (36 m) tall occupy rocky ridges and moraines, where the ground is covered with a rich growth of mosses. *Usnea* lichens may be draped over heath plants. On boggy ground, fescue grass and carex sedge tussocks dominate, with lady's mantle growing between the clumps. Because the tussocks cover most of the ground, the general appearance is grassland with occasional giant rosettes of *Lobelia keniensis*.

The treeless environment can be divided into lower alpine, upper alpine, and nival zones, based on dominance of different giant rosettes, even though considerable overlap of species and interdigitation of communities occur, depending on slope, aspect, drainage, and other local characteristics. (Zones are more precisely defined on mountains not subjected to glaciation, such as Mount Kilimanjaro, where aspect and slope are less variable.)

The lower alpine zone, roughly 12,000 ft (3,650 m), is marked by the abundance of the ground-level giant senecios (*Senecio brassica*) and lack of the tree-like giant groundsels (*Dendrosenecio keniodendron*). Two basic habitats, meadows and ridgetops, are encountered. The flat or gently sloping meadows and valley bottoms are usually wet and characterized by tussock grasslands, an almost continuous cover of fescue with few or no shrubs. Fescue grass forms clumps to more than 3 ft (1 m) high, and smaller carex sedge tussocks also occur. Mosses and smaller rosettes grow in the dark humus soil between tussocks. Scattered groups of lobelia rosettes, 1 ft (0.3 m) high and 18 in (45 cm) in diameter, occupy semibogs where soil is rich in humus and kept damp by seepage or standing water. Tussocks of nodding hairgrass with patches of prostrate senecios grow where meadows border rocky areas.

Weathered and eroded ridgetops and sometimes old eroded moraines offer a rockier, more exposed, and drier habitat with plant communities that continue into the upper alpine zone.

A cover of nodding hairgrass and other grasses extends up to 15,000 ft (4,500 m), the grasses becoming shorter with increased exposure. Few rosettes grow between the grass tussocks because drainage is excessive and soils become dry, and frost-heaving is too severe for soils to remain stable. Creeping plants and woody plants such as strawflower are found only in sheltered spots. Scattered giant groundsels, more common at higher elevations, grow on well-drained ridges, indicating a somewhat harsher environment and transition into the upper alpine zone. Dung accumulations near hyrax burrows support dense growth of an endemic succulent.

Characterized by the appearance of giant groundsels in distinct communities, the upper alpine zone lies at elevations of about 14,000 ft (4,250 m). Valley walls, the steeply sloping sides of glaciated valleys, where soil is damp but well-drained, are typically covered by a pattern of a groundsel and a senecio, according to moisture requirement. Higher on the slope, tall tree-like giant groundsels are found on the well-drained, shallow soils with access to abundant subsurface water, while the prostrate giant senecios grow lower down the valley wall on waterlogged finer soils and around lakes. Thick masses of lady's mantle shrubs grow at the base of the giant groundsels. Protected areas enclosed by steep cliffs allow heath vegetation to grow at higher-than-usual elevations. Solifluction terraces and damp, flat ground on valley floors support fescue or bentgrass tussocks. Ridgetops are plagued by wind and intensive frost. Sparse grass tussocks exist up to 14,500 ft (4,400 m) on smooth ridges, but almost none occur over 15,000 ft (4,500 m). Lady's mantle stands occupy ground just below the ridgetops, along with woody strawflowers. Rocky outcrops are more varied, but include herbaceous senecios in the lee of boulders in windy areas. The small *Senecio purtschelleri* is common around hyrax colonies.

Rather than permanent snow and ice, the nival zone on Mount Kenya refers to elevations above 15,000 ft (4,500 m) from which glaciers have recently retreated. Stunted plants grow in sheltered locations, and pioneering species such as herbaceous senecios (*Senecio keniophytum* and *S. purtschelleri*) can be found just 25 ft (7.5 m) from the ice. Brown's strawflower, a hardy shrub and the highest plant on the mountain, is found sheltering in rock cracks with little soil at elevations up to 16,000 ft (4,875 m) that have long been ice-free.

··············································································································

***East African alpine animals.*** Equatorial East African mountains are not well inventoried, but the more conspicuous species are known. Mammalian herbivores on Mount Kenya include rock hyrax, groove-toothed rat, and the common duiker. Moraines or crags adjacent to tussock grassland and water constitute their habitat, each animal occupying a microenvironment. Hyrax live among the rocks, rats dig burrows at the bases of giant senecios or grass tussocks, and duikers prefer giant groundsel forest. Hyrax is commonly seen in the afroalpine, with different subspecies on different mountains (see Figure 4.12). They eat mosses and grasses, types varying according to proximity to their burrows. Fur on the mountain hyrax is over 2 in (5 cm) long compared with only 0.5 in (1.3 cm) on lowland savanna species, and the alpine form has a larger body size. Also unlike the lowland hyrax, they drink water, often trampling trails to lakes or streams. Although their burrows are just deep holes between rocks, temperatures inside are above freezing and

**Figure 4.12** The herbivorous hyrax lives in burrows between rocks in the afroalpine. *(Courtesy of Shutterstock. Copyright: Steffen Foerster Photography.)*

considerably warmer than the outside air. Plants growing at burrow entrances are not eaten, which helps to conceal burrow locations.

Long-furred groove-toothed rats are commonly found on tussock grassland. They burrow into the base of giant groundsels and then up into the plant as far as the base of the leaves where temperatures remain more constant. The rats feed on plant roots and seeds. Fewer in number, duikers live in the heath zone or in giant groundsel forest where they browse on lady's mantle and other woody plants. Other small mammals include a shrew, the harsh-furred mouse, and giant mole rat.

Predators, feeding mainly on hyrax, include an occasional leopard, wild dog, and several raptors such as Mackinder's Owl, Augur Buzzard, Verreaux Eagle, and Lammergeier.

The three most common birds on Mount Kenya are the Scarlet-tufted Malachite Sunbird, the Hill Chat, and Streaky Seed Eater; all live in leaf girdles of giant groundsels. The sunbird, which breeds all year in the alpine zone, takes good advantage of giant rosettes. It uses its long bill to get nectar from deep inside lobelia flowers and also feeds on insects, especially flies that live in the inflorescences. It also eats midges that breed in the pools of water that collect in lobelia rosettes, and strips the hairy covering from senecio leaves to line its nest. The Hill Chat, common in both heath and alpine environments, feeds mainly on carabid beetles, weevils, and spiders from the ground and from the leaves of senecios and lobelias. It does not, however, touch insects on the lobelia inflorescences used by the sunbird. The seed eater of course eats seeds.

Insects also take advantage of the microclimates in plants. Two moths living in fescue tussocks build silk tubes from the base to the outer leaves. The insects move through the tube in response to diurnal temperature fluctuations, spending the hottest part of the day and cold nights at the base of the plant and only venturing to the outer leaves in early morning and evening when air temperatures are more moderate. Insects take shelter from the cold in other animal nests, under rocks, rosettes of lobelias or senecios, or in dead leaves on giant groundsel stems. Midges breed in lobelia rosettes and shelter in giant groundsels. Many specialized insects, especially beetles, are restricted to the environment of mole rat burrows. Some species are apparently limited to one mountain or habitat, but there has been little study of afroalpine invertebrates to confirm this. However, many species have evolved; in one small area on Mount Elgon, 22 different species and subspecies of the carabid beetle genus *Trechus* were counted.

The only resident reptile is the alpine meadow lizard, which lives in tussocks and under stones.

## Ethiopian Montane Moorlands

The Ethiopian plateau is a large expanse of land more than 6,500 ft (2,000 m) in elevation with mountain chains and peaks more than 9,850 ft (3,000 m) high. Although the Simen Mountains in the north are the highest, with Ras Dejen at 15,158 ft (4,620 m), the Bale Mountains in the south make up the largest continuous afroalpine area. In contrast to the mountains of East Africa, which are surrounded by wetter climates, the Ethiopian highlands rise from dry lands. The mountain climates are largely unknown, but change according to elevation. Annual rainfall varies considerably, from 100 in (2,500 mm) in the southwest to 40 in (1,000 mm) in the north. Depending on locale, the dry season can be as short as two months or as long as 10. Frost occurs all year, but especially in winter from November to March. Diurnal temperature ranges can be extreme, from 5° F (−15° C) at night to 79° F (26° C) during the day.

About 80% of the land above 9,850 ft (3,000 m) in Africa is in Ethiopia. The lower part of the alpine zone is a montane forest or grassland now degraded by grazing. Several alpine lakes add interest to the landscape. The flora here has affinities with both Palaearctic and Afrotropical plant realms, and high levels of endemism are found in this isolated volcanic upland. The Bale Mountains are a center of endemism, with 20% of their wild mammals unique to the range. Development of the African Rift Valley divided the plateau and its biota into two parts. The highest areas were glaciated during the Pleistocene and were uncovered and colonized only a few thousand years ago.

Adaptations to the high-altitude environment include gigantism, such as giant lobelias, tree heather, and giant St. John's wort. Many St. John's worts are herbaceous or shrubby, but in Ethiopia, the species grows up to 40 ft (12 m) tall with a

**Figure 4.13** St. John's wort grows to unusually large sizes in the Ethiopian highlands. *(Courtesy of Rainer W. Bussmann, Missouri Botanical Garden.)*

trunk 10 in (25 cm) in diameter (see Figure 4.13). Some adaptations serve to reduce transpiration. Leaves of lobelias are thick and leathery, and many species have a small leaf surface area. The dry, paper-like flowers on many perennial plants are able to withstand harsh winds.

Vegetation is predominantly a grassland with shrubs or a moorland dominated by strawflower scrub and tussock grasses. The dominant treeline scrub consists of *Philippia,* tree heather, and other shrubs. Smaller plants in the primarily bare soil include strawflower, lady's mantle, Junegrass, and hairgrass. Wetter areas are dominated by carex sedges. Above 12,150 ft (3,700 m), carex sedge and fescue grasslands with giant lobelias dominate. A single species of lobelia is the only giant rosette in Ethiopia. In the highest regions, a fescue and straw-flower scrub community extends all the way to mountain summits, although high-elevation cliffs and rocky slopes have little vegetation. The Ethiopian highlands differ from other afroalpine mountains in that they have few shrubby lady's mantles and no giant groundsels. They also have African rose and yellow primrose, typical genera of Europe or Asian alpine areas. Lichens often cover woody trees or shrubs.

The Ethiopian highlands is a unique area in terms of animals and has several endemic and threatened mammals. Simen National Park was named a World Heritage Site in 1978. Significant is the extremely rare Ethiopian wolf, also called the Simen fox. It lives on the open moorlands above 9,800 ft (3,000 m) and eats burrowing rodents such as the giant mole rat. Evolutionarily related to the European grey wolf, it provides a major example of prior connections with Eurasia. The giant mole rat, a large endemic rodent, also has a European origin. Other rodents in the wolf's diet include diurnally active grass rats, and Starck's hare. Speciation among rodents has been high. Other rodents endemic to the Ethiopian alpine are the giant climbing mouse, narrow-headed rat, and black-clawed brush-furred rat. Near-endemic animals include Walia ibex, mountain nyala, and gelada baboon. The ibex, nyala, and baboon, along with klipspringer and rock hyrax found in rocky habitats, point to relationships with desert ecosystems.

Alpine lakes and streams provide excellent avian habitat, particularly to over-wintering Palaearctic birds. Ducks that can be seen in the thousands include wigeon and shovelers. Wading birds such as Ruffs, and Greenshanks are also abundant. Golden Eagle, Chough, and Ruddy Shelduck, all Palaearctic species, breed in the Ethiopian alpine, their only known breeding site outside the Northern Hemisphere temperate zone.

Good rainfall has long attracted agricultural peoples to the Ethiopian Plateau and continues to foster increasing human populations. The Ethiopian highlands have been cultivated for centuries and much of the area is degraded. The more in-hospitable environments of most of the afroalpine zone have made it easier to pre-serve. The montane and heath zones in the mountains of Equatorial East Africa, however, are affected by human activities.

## Oceania: Hawaiian Islands

Some 2,500 mi (4,000 km) from North America and 2,000 mi (3,200 km) from the nearest islands of comparable height in French Polynesia, Hawaii is one of the most isolated groups of islands in the world. Mauna Kea, on the big island of Hawaii, is highest at 13,802 ft (4,207 m), closely followed by nearby Mauna Loa at 13,681 ft (4,170 m). Haleakala on Maui is only 10,026 ft (3,056 m). Because these volcanoes built up from the seafloor and have never been connected with a continent, many continental lifeforms such as ants, coniferous trees, most land birds, and all mam-mals except a bat, were unable to reach the islands. Long-distance dispersal, by means of wind, rafting on the ocean, or attached to feathers or feet of birds was infre-quent, and plants or animals that did arrive evolved in isolation. The diversity of Hawaii's biota stems from adaptive radiation, meaning that immigrant plants and animals evolved into many different species to occupy vacant niches or habitats. Polynesians colonized the islands about 300 years ago, bringing with them pigs, jungle fowl, dogs, and Polynesian rats, and (unintentionally) geckos, skinks, and snails. Many food plants such as taro were also introduced. Cattle, goats, and sheep were introduced by Europeans at a later date.

Even though Hawaii lies at or near tropical latitudes (20° N), summits of the higher mountains are cold. Temperatures on Mauna Kea average 39° F (4° C), and the mountain was glaciated in the Pleistocene. Snow frequently falls on Mauna Kea and Mauna Loa at elevations as low as 8,200 ft (2,500 m), but only occasionally on Haleakala. Fresh lava or areas of cinders and ash have seen little soil development.

Because Hawaii is under the influence of both the trade winds and subtropical high pressure, an inversion layer develops. Warm moist trade wind air becomes cooler as it rises, while the descending air in the high-pressure cell becomes warmer as it sinks. This results in warm air overlying cool air at an altitude between 5,000 and 7,000 ft (1,500–2,150 m), called an inversion, because the normal trend of temperatures becoming cooler with elevation is turned upside down or inverted. The inversion layer prevents trade wind air from rising any higher, blocking moisture from the alpine zone. Haleakala's summit gets less than 30 in (750 mm) of precipitation annually, and Mauna Kea is even drier, with less than 15 in (380 mm). The dry, clear skies on Mauna Kea provide ideal conditions for astronomical observations, and one of the world's major observatories sits on its summit.

Closed forest extends up to about 6,500 ft (2,000 m), which coincides with the level of trade wind inversions. Treeline generally occurs about 9,000 ft (2,750 m) and is characterized by shrublands or open parkland with small ohia trees. The change from forest to alpine scrub above treeline is abrupt due to the inversion's effect on precipitation. Not only is the alpine zone much drier than the lowlands, it also receives intense radiation due to the absence of cloud cover.

Three sparsely vegetated life zones occur above treeline: alpine scrub, alpine desert, and the aeolian zone. Around 8,200 ft (2,500 m) elevation, an alpine scrubland dominated by prostrate shrubs of pukeawe, Hawaiian blueberry, mamame, pilo, cranesbill, and tarweed typifies the change from wet forest to dry subalpine. The lower stature of plants is caused by low precipitation at this elevation, most of it falling as winter snow. Nightly frost is common. On drier slopes the alpine scrub extends to lower elevations. At 8,000 ft (2,450 m) on the northeastern side of Haleakala, an open grassland with endemic hairgrass bunches dominates. At the highest levels of plant growth, at about 11,300 ft (3,450 m) and just before the alpine desert, prostrate shrubs of pukeawe and Hawaiian blueberry grow with several bunchgrasses, such as hairgrass, panicum, and bentgrass, carex sedges, and ferns. Two conspicuous alien (nonnative) species in the alpine scrub are gosmore, a plant that resembles a dandelion, and wooly mullein. Except for a few hardy individuals, the alpine scrub level is the zone where silverswords, the giant rosettes of Hawaii, grow.

Above the alpine scrub zone to the upper limit of plant life at about 11,300 ft (3,450 m) is an alpine desert of fresh ash, cinders, and lava flows extending from the summits of Mauna Kea, Mauna Loa, and Haleakala. Frost occurs almost every night. Permanent ice can be found less than 3.3 ft (1 m) beneath the cinder surface on the summit cones of Mauna Kea, and Haleakala has some active patterned ground such as stone stripes. Plant life is limited to lichens and mosses.

The highest part of the mountains, above the alpine desert, is the aeolian zone. Life is limited to native arthropods adapted to the extreme cold and barren environment and surviving on plant and animal material carried on upslope winds. Caterpillars of a flightless moth on Haleakala survive by eating leaves trapped in their webs. At least one spider eats insects blown upslope. The Mauna Kea wēkiu bug and its close relative the Mauna Loa wēkiu bug are also flightless. Instead of a vegetable diet like their low-elevation relatives, they suck the body fluids of cold or dead insects. A change in their blood that prevents freezing in spite of low temperatures is so powerful that the bugs are extremely sensitive to heat, even that of a human hand.

Most other animal life in the Hawaiian alpine consists of introduced animals that have obliterated native species. Pigs, rats, dogs, and jungle fowl dominate. Both grassland and shrubland in Haleakala National Park incurred considerable damage from pigs before the 1980s when they were eliminated from the park. While digging for preferred food of gosmore and bracken fern, they disrupted native plants, which allowed exotics to spread and displace even more natives. Goats, sheep, and cattle were also detrimental to native vegetation.

**Silverswords.** Silverswords (`ahinahina in Hawaiian), endemic to the islands of Hawaii and Maui, are rosettes of stiff succulent leaves that grow in the alpine scrub regions of Mauna Kea, Mauna Loa, and Haleakala (see Plate XVI). Although they appear silvery white, the long, pointed leaves are actually lime green but densely covered with silky silver hairs. Silversword habitat receives some of the highest solar radiation ever measured, and the silver hairs, easily damaged by touching, reflect both visible and ultraviolet light.

Silverswords and greenswords are among three genera and 30 species in a group of related plants called the Silversword Alliance that includes herbs, vines, shrubs, trees, and rosette plants, all of which are endemic to Hawaii. The alliance is a good example of adaptive radiation and convergent evolution because all the species evolved from a North American tarweed in the sunflower family that arrived in Hawaii about 5.2 million years ago. The taxonomy of silverswords is in flux, with questions as to whether some are species or subspecies.

Of the 23 species of Hawaiian tarweeds, only some with succulent leaves grow in the alpine zone. The three species of silverswords and two species of Maui greenswords are all in the genus *Argyroxiphium*. They only grow in localized distributions above 4,900 ft (1,500 m) on the islands of Maui and Hawaii. Two silverswords grow in the alpine scrub of Haleakala, Mauna Kea, and Mauna Loa, and like desert plants, they are adapted to intense sun and dry conditions. The third is eke (*A. caliginis*), a tiny silversword only 6 in (15 cm) in diameter with a 1.5 ft (0.5 m) flower stalk, which occupies bogs at the transition between the forest and alpine zones in western Maui. The two Maui greenswords have few silver hairs on their green leaves and grow in misty sites. Two greenswords on Kauai belong to a different genus (*Wilkesia*), and are known as iliau in Hawaiian. They are endemic to that island and do not grow in an alpine environment.

Leaves of silverswords are long and narrow, 6–16 in (15–40 cm long) and 0.2–0.6 in (0.5–1.5 cm) wide. The genus has evolved a growthform similar to the giant rosettes in tropical Africa and in the tropical Andes, a rosette that protects the interior growing point and from which a tall flower stalk emerges. Water and nutrients stored in the leaves are used to produce the tall inflorescence. The plant grows as a ground-hugging rosette or with a very short stem. Two subspecies vary in the shape of the inflorescence and geographic location: one is found in Haleakala crater, the other on Mauna Kea. Both occupy dry alpine cinder habitats between 7,000 ft (2,150 m) and 12,300 ft (3,750 m) on the volcanoes. A third species, Mauna Loa silversword, is limited to that volcano and its locale is protected in Hawaii Volcanoes National Park. A short, shrubby tarweed found on exposed windy slopes hybridizes freely with silverswords in nature. More than 30 identified crosses indicate their recent evolution.

A silversword flowers only once in its lifetime. It takes 15–50 years for the plant to mature from seed to flowering stage. It remains a compact rosette up to 2 ft (0.6 m) in diameter until it sends up a flower stalk. Flowering occurs mid-June to November, and after the seed ripens three months later, the plant dies. The inflorescence, which carries 50–600 pink, white, or red daisy-like flowers, grows up to 10 ft (3 m tall) and 30 in (75 cm) wide. Each plant produces thousands of seeds. Because they require cross-pollination, all plants in an area generally flower at the same time. Sticky hairs on its leaves, stems, and central stalk trap crawling insects and prevent them from accessing the pollen, allowing flying pollinators such the native yellow-faced bee, to carry pollen to other plants. Occasionally small rosettes will branch out from the main body, a benefit because only the rosette with the flower stalk will die.

Early in the twentieth century, silversword populations were highly endangered because of vandalism and browsing by goats and cattle. Root systems can be easily crushed in loose cinders. After vandalism was curtailed and cattle were eliminated from Haleakala National Park in the 1930s and goats in the 1980s, the population came back from a low of 2,000–4,000 plants to more than 60,000 now.

## Further Readings

Coe, Malcolm James. 1967. *The Ecology of the Alpine Zone of Mount Kenya.* The Hague: Dr. W. Junk Publishers.

Line, Les. December 2003–January 2004. "Farewell to Flamingos?" *National Wildlife Magazine.* http://www.nwf.org/nationalwildlife/article.cfm?issueID=65&articleID=875.

Palomar College. n.d. The Silversword Alliance. http://waynesword.palomar.edu/ww0903b.htm.

Public Broadcasting Service. n.d. "Salt Flat Living." Andes: The Dragon's Back. http://www.pbs.org/wnet/nature/andes/saltflat.html.

Wheeler, Jane C. n.d. "Evolution and Present Situation of the South American Camelidae." http://www.conopa.org/camelidos/historia.php.

# Appendix

## Biota of Tropical Alpine Tundra Biome (arranged geographically)

### South American Alpine Tundra

#### Some Characteristic Plants of the Andean Páramo

*Trees and treeline*

| | |
|---|---|
| Polylepis (in the Rose family) | *Polylepis sericea* |
| Chachacomo | *Escalonia* spp. |

*Shrubs*

| | |
|---|---|
| Chuquiragua or Flower of the Andes | *Chuquiragua* spp. |
| Jata or Candlebush | *Locicaria ferruginea* |
| In the Mahogany family | *Schmardaea microphylla* |
| Quinine or Cedro colorado | *Chichona officinallis* |
| St. John's wort | *Hypericum laricifolium* |

*Graminoids*

| | |
|---|---|
| Reedgrass | *Calamagrostis* spp. |
| Fescue grass | *Festuca* spp. |
| Dwarf bamboo | *Chusquea* spp. |
| Bentgrass | *Agrostis* spp. |
| Brome grass | *Bromus catharticus* |
| Needlegrass | *Stipa* spp. |

*Forbs*

*Giant rosettes*

| | |
|---|---|
| Large espeletia | *Espeletia timotensis* |
| Small espeletia | *Espeletia moritziana* |
| Midsize espeletia | *Espeletia spicata* |
| Prostrate stem espeletia | *Espeletia semiglobulata* |
| Short espeletia | *Espeletia schultzii* |
| Wooly espeletia | *Espeletia hartwegiana* |

| Puya | *Puya clava-herculis* |
| Puya | *Puya raimondii* |
| Puya | *Puya hamata* |

*Cushions*

| Azorella (in the Aralia family) | *Azorella crenata* |
| Azorella (in the Aralia family) | *Azorella julianii* |
| Sandwort | *Arenaria* spp. |
| Ragwort | *Senecio* spp. |

*Leafy*

| Iceland purslane | *Koeniga islandica* |
| Geranium | *Geranium* spp. |
| Gentian | *Gentiana* spp. |
| Paintbrush | *Castilleja* spp. |
| Buttercup | *Ranunculus* spp. |

**Cryptogams**

| Sphagnum moss | *Sphagnum* |

## Some Characteristic Animals of the Andean Páramo

**Herbivores**

| Mountain tapir | *Tapirus pinchaque* |
| Little red brocket deer | *Mazama rufina* |
| Northern pudu | *Pudu mephistophiles* |
| Rice rat | *Microryzomys minutus* |
| Páramo rabbit | *Sylvilagus brasiliensis.* |
| White-tailed deer | *Odocoileus virginianus* |

**Carnivores**

| Thomas' small-eared shrew | *Cryptotis thomasi* |
| White-eared opposum | *Didelphis albiventris.* |
| Long-tailed weasel | *Mustela frenata* |
| Páramo wildcat | *Felis tigrina* |
| Puma | *Puma concolor* |

**Birds**

| Andean Condor | *Vultur gryphus* |
| Páramo Pipit | *Anthus bogotensis* |
| Streak-backed Canastero | *Asthenes wyatti* |
| Merida Wren | *Cistothorus meridae.* |
| Sedge Wren | *Cistothorus platensis* |
| Black-chested Buzzard Eagle | *Geranoactus melanoleucus* |
| White-rumped Hawk | *Buteo leucorrhous* |

## Some Characteristic Plants of the Andean Puna

### Trees and treeline
| | |
|---|---|
| Butterfly bush or Colle | *Buddleja coriaceae* |
| Polylepis or Quenoa (in the Rose family) | *Polylepis tomentella* |
| Polylepis or Quenoa (in the Rose family) | *Polylepis tarapacana* |

### Shrubs
| | |
|---|---|
| In the Pea family | *Adesmia* spp |
| Broom rape | *Baccharis incarum* |
| Broom rape or Chijua | *Baccharis boliviensis* |
| Tolillar | *Fabiana densa* |
| Atacama saltbush | *Atriplex atacamensis* |

### Graminoids
| | |
|---|---|
| Carex sedge | *Carex* spp. |
| Sedge | *Oreobolus* spp. |
| Needlegrass | *Stipa jehu* |
| Peruvian feathergrass or Ichu | *Stipa ichu* |
| Fescue grass | *Festuca* spp. |
| Reedgrass | *Calamagrostis* spp. |
| Bentgrass | *Agrostis* spp. |
| Mountain bamboo | *Chusquea* spp. |
| Pampas grass | *Cortaderia* spp. |
| Saltgrass | *Distichlis humilis* |
| Rush | *Juncus* spp. |
| Bulrush | *Scirpus* spp. |
| Cushion rush | *Distichia muscoides* |
| Cushion rush | *Oxychloe andina* |

### Cacti
| | |
|---|---|
| Old man of the Andes | *Oreocereus* spp. |
| San Pedro cactus | *Trichocereus pachanoi* |

### Forbs
*Giant rosettes*
| | |
|---|---|
| Puya | *Puya raimondii* |
| Puya | *Puya hamata* |

*Cushions*
| | |
|---|---|
| Azorella (in the Aralia family) | *Azorella yarita* |
| Azorella (in the Aralia family) | *Azorella compacta* |
| Cushion plantain | *Plantago rigida* |

*Leafy*
| | |
|---|---|
| Gosmore | *Hypochoeris* spp. |
| In the Rose family | *Lachemilla* spp. |
| In the Sunflower family | *Culcitium* spp. |

| Pickleweed | *Salicornia pulvinata* |
| Seepweed | *Suaeda foliosa* |

## Some Characteristic Animals of the Andean Puna

### Herbivores
| Vicuna | *Vicugna vicugna* |
| Guanaco | *Lama guanicoe* |
| Llama | *Lama glama* |
| Alpaca | *Lama pacos* |
| Chinchilla | *Chinchilla brevicaudata*[b] |
| Golden vizcacha rat | *Pipanococtomys aureus*[a] |

### Carnivores
| Puma | *Felis concolor* |
| Andean fox | *Pseudalopex culpaeus* |
| Andean Mountain cat | *Oreailurus jacobita*[b] |
| Andean hairy armadillo | *Caetophractus nationi* |

### Birds
| Ash-breasted Tit-tyrant | *Anairetes alpinus*[a] |
| Royal Cinclodes | *Cinclodes aricomae*[a] |
| Berlepsch's Canastero | *Asthenes berlepschi*[a] |
| Line-fronted Canastero | *Asthenes urubambensis*[a] |
| Olivaceaous Thornbill | *Chalcostigma olivaceum*[a] |
| Scribble-tailed Canastero | *Asthenes maculicauda*[a] |
| Gray-bellied Flower Piercer | *Diglosa carbonaria*[a] |
| Darwin's Rhea or Suri | *Pterocnemia pennata* |
| Puna Tinamou | *Tinamotis pentlandii* |
| James Flamingo | *Phoenicopterus jamesi*[b] |
| Andean Flamingo | *Phoenicopterus andinus*[b] |
| Chilean Flamingo | *Phoenicopterus chilensis* |

*Notes:* [a]Endemic; [b]Rare and endangered.

## African Alpine Environments

## Some Characteristic Plants of Equatorial East Africa

### Treeline
| Bamboo | *Arundinaria alpina* |
| Protea | *Protea kilimandscharica* |
| Tree heather | *Erica arborea* |
| Philippia heath | *Philippia excelsa* |
| Philippia heath | *Philippia keniensis* |

(*Continued*)

**Low or subshrubs**

| | |
|---|---|
| Strawflower or Everlasting | *Helichrysum stublmannii* |
| Kilimanjaro strawflower | *Helichrysum kilimanjari* |
| Johnston's lady's mantle | *Alchemilla johnstonii* |
| Lady's mantle | *Alchemilla argyrophylla* |
| Brown's strawflower or Everlasting | *Helichrysum brownei* |

**Graminoids**

| | |
|---|---|
| Carex sedge | *Carex monostachya* |
| Fescue grass | *Festuca pilgeri* |
| Bentgrass | *Agrostis trachyphylla* |
| Bluestem grass | *Andropogon* spp. |
| Hairgrass | *Koeleria* spp. |
| Pentaschistis grass | *Pentaschistis minor* |
| Nodding hairgrass | *Deschampsia flexuosa* |

**Forbs**

*Giant rosettes*

| | |
|---|---|
| Giant groundsel | *Dendrosenecio keniodendron* |
| Giant groundsel | *Dendrosenecio johnstonii* |
| Giant groundsel | *Dendrosenecio brassiciformis* |
| Giant groundsel | *Dendrosenecio keniensis* |
| Giant lobelia | *Lobelia telekii* |
| Giant lobelia | *Lobelia keniensis* |
| Giant senecio or Giant groundsel | *Senecio brassica* |
| Giant senecio or Giant groundsel | *Senecio battescombie* |

*Leafy*

| | |
|---|---|
| Lady's mantle | *Alchemilla cyclophylla* |
| Sedum (succulent) | *Sedum ruwenzoriense* |
| Senecio | *Senecio purtschelleri* |
| Senecio | *Senecio keniophytum* |

**Cryptogams**

| | |
|---|---|
| Usnea (Fruticose lichen) | *Usnea* |

## Some Characteristic Animals of Equatorial East Africa

**Herbivores**

| | |
|---|---|
| Rock hyrax | *Procavia johnstoni* |
| Groove-toothed rat | *Otomys orestes* |
| Common duiker | *Sylvicapra grimmia altivallis* |
| Harsh-furred mouse | *Laphuromys a. aquilas* |
| Giant mole rat | *Tachyryctes rex* |

***Carnivores***

| | |
|---|---|
| Leopard | *Felis pardus* |
| Shrew | *Crocidura alex alpina* |
| Wild dog | *Lycaon pictus lupinus* |

***Birds***

| | |
|---|---|
| Mackinder's Owl | *Bubo capensis* |
| Augur Buzzard | *Buteo rufofuscus augur* |
| Verreaux Eagle | *Aquila verreauxii* |
| Lammergeier or Bearded Vulture | *Gyptaetus barbatus* |
| Scarlet-tufted Malachite Sunbird | *Nectarinia johnstoni johnstoni* |
| Hill Chat | *Pinarochroa sordida ernesti* |
| Streaky Seed Eater | *Serinus striolatus stridatus* |

***Reptiles***

| | |
|---|---|
| Alpine meadow lizard | *Algyroides alleni* |

## Some Characteristic Plants of the Ethiopian Plateau

***Treeline***

| | |
|---|---|
| Tree heather | *Erica arborea* |
| Heath | *Erica trimera* |
| Giant St. John's wort | *Hypericum revolutum* |
| Philippia heath | *Philippia* spp. |

***Low shrubs***

| | |
|---|---|
| Lady's mantle | *Alchemilla haumannii* |
| Strawflower | *Helichrysum citrispinum* |

***Graminoids***

| | |
|---|---|
| Carex sedge | *Carex monostachya* |
| Junegrass | *Koeleria* spp. |
| Hairgrass | *Aira* spp. |
| Fescue grass | *Festuca macrophylla* |
| Fescue grass | *Festuca abyssinica* |

***Forbs***

*Giant rosettes*

| | |
|---|---|
| Giant lobelia | *Lobelia rynchopetalum* |

*Leafy*

| | |
|---|---|
| African rose | *Rosa abyssinica* |
| Yellow primrose | *Primula verticillata* |

***Cryptogams***

| | |
|---|---|
| Usnea (Fruticose Lichen) | *Usnea* spp. |

## Some Characteristic Animals of the Ethiopian Plateau

### Herbivores

| | |
|---|---|
| Walia ibex | *Capra ibex walie* |
| Mountain nyali | *Tragelaphus buxtoni* |
| Klipspringer | *Oreotragus oreotragus* |
| Giant mole rat | *Tachyoryctes macrocephalus*[a] |
| Grass rat | *Lophuromys melanonyx* |
| Stark's hare | *Lepus starcki* |
| Giant climbing mouse | *Megadendromus nikolausi*[a] |
| Narrow-headed rat | *Stenocephalemys albocaudata*[a] |
| Black-clawed brush-furred rat | *Lophuromys melanoyx*[a] |
| Rock hyrax | *Procavia capensis* |

### Carnivores

| | |
|---|---|
| Ethiopian wolf or Simen fox | *Canis simensis*[a] |
| Gelada baboon | *Theropithecus gelada* |

### Birds

| | |
|---|---|
| Widgeon | *Anas penelope* |
| Shoveler | *Anas clypeata* |
| Ruff | *Philomachus pugnax* |
| Greenshanks | *Tringa nebularia* |
| Golden Eagle | *Aquila chrysaetos*[b] |
| Chough | *Pyrrhocorax pyrrhocorax*[b] |
| Ruddy Shelduck | *Tadorna ferruginea*[b] |

*Notes:* [a]Endemic; [b]Migratory breeder.

# Oceania Alpine Tundra

## Some Characteristic Plants of Hawaii Alpine

### Treeline

| | |
|---|---|
| Ohia (in the Myrtle family) | *Metrosideros polymorpha* |

### Low or subshrubs

| | |
|---|---|
| Pukeawe (in the Heath family) | *Styphelia tameiameiae* |
| Hawaiian blueberry | *Vaccinium reticulatum* |
| Mamame (in the Pea family) | *Sophora chrysophylla* |
| Pilo (in the Madder family) | *Coprosma montana* |
| Cranesbill | *Geranium cuneatum* |
| Tarweed or Naènaè (in the Sunflower family) | *Dubautia menziesii.* |

### Graminoids

| | |
|---|---|
| Hairgrass | *Deschampsia nubigena* |
| Panicum grass | *Panicum* spp. |
| Bentgrass | *Agrostis sandwicensis* |

*Forbs*
*Giant rosettes*

| | |
|---|---|
| Haleakala silversword | *Argyroxiphium sandwicense* ssp. *macrocephalum* |
| Mauna Kea silversword | *Argyroxiphium sandwicense* ssp. *sandwicense* |
| Mauna Loa silversword | *Argyroxiphium kauense* |
| Maui greensword | *Argyroxiphium kai* |
| Maui greensword | *Argyroxiphium virescens* |
| Eke | *Argyroxiphium caliginis* |

*Leafy*

| | |
|---|---|
| Gosmore or Hairy cat's ear | *Hypochoeris radicata*[a] |
| Wooly mullein | *Verbascum thapsus*[a] |
| Bracken fern | *Pteridium aquilinum* |
| Fern | *Asplenium* spp. |

*Note:* [a]Introduced.

## Some Characteristic Animals of Hawaii Alpine

**Herbivores**

| | |
|---|---|
| Pig | *Sus scrofa*[a] |
| Polynesian rat | *Rattus exulans*[a] |

**Carnivores**

| | |
|---|---|
| Dog | Canis familiaris[a] |

**Birds**

| | |
|---|---|
| Jungle Fowl | *Gallus gallus*[a] |

**Insects**

| | |
|---|---|
| Haleakala flightless moth | *Thryocopa apatela* |
| Mauna Kea wēkiu bug | *Nysius wekiuicola* |
| Mauna Loa wēkiu bug | *Nysius aa* |

*Note:* [a]Introduced.

# Glossary

**Ablation.** Means by which ice is lost from a glacier, including melting, calving, and sublimation.

**Aeolian.** The zone on mountains above the limit of plant life, where wind-blown debris and insects may accumulate.

**Albedo.** Reflectivity of a surface, refers to how much solar radiation is reflected.

**Alpine.** The zone on mountains above the region of trees.

**Alpine Glacier.** Ice that moves downslope in mountain valleys. Also called valley glaciers because the ice is confined to the valley and does not cover the entire landscape.

**Annual.** A plant that completes its life cycle in one year or one growing season.

**Antarctic.** Strictly, the latitudes between the Antarctic Circle at $66^1/_2°$ S and the South Pole. Can refer to the general region near or south of the Antarctic Circle.

**Arctic.** Strictly, the latitudes between the Arctic Circle at $66^1/_2°$ N and the North Pole. Can refer to the general region near or north of the Arctic Circle.

**Arete.** A narrow, sharp-edged ridge between two valley glaciers or glaciated valleys.

**Aspect.** Direction toward which a slope faces. Also known as exposure.

**Behavioral.** Activities or behavior of animals.

**Biome.** A large region with similar vegetation, animal life, and environmental conditions.

**Biota.** The combined flora and fauna, including all the plants and animals.

**Boulder Clay.** Unsorted jumble of rocks and smaller particles deposited by a glacier.

**Bryophyte.** Group of plants that includes mosses.

**Bulbil.** Small plant-like shoot produced on a flower stalk.

**Calving.** Large pieces of ice breaking off a glacier.

**CAM (Crassulacean Acid Metabolism).** A form of photosynthesis used by plants to conserve water.

**Canopy.** The uppermost layer of foliage in vegetation.

**Cenozoic.** A recent geologic time period, roughly 65 million years ago to the present.

**Chamaephytes (Raunkiaer).** Plants that hold their regenerating buds just above soil level.

**Circumpolar.** Distribution that encircles the North Pole.

**Climate.** Typical weather (especially temperature and precipitation) patterns during a normal year that are experienced over decades or centuries. Weather refers to the conditions of the atmosphere at any given moment.

**Cogener.** Plants in the same genus.

**Community.** The plants and animals assembled in a given area. Sometimes refers to only a subset of these organisms, such as the plant community or the bird community.

**Conifer.** A cone-bearing plant such as pine and spruce.

**Continental Glacier.** Ice that covers a large part of a continent, such as Greenland. Also called ice sheet.

**Continentality.** The effect a large land mass has on seasonal temperature variations. Continental areas are warmer in summer and colder in winter. Compare with Maritime.

**Cover.** The proportion of a surface on which vegetation occurs, usually measured as a percent.

**Cretaceous.** A geological time period from roughly 145 million years ago to 65 million years ago.

**Crustose (lichen).** Crustlike. Compare with foliose and fruticose.

**Cryptogam.** Plant group that includes algae, fungi, mosses, lichens, and ferns, which lack flowers. Most are nonvascular plants and have no differentiated tissue to transport water or nutrients.

**Cushion Plant.** A low-growing, multistemmed plant that grows as a dense mound.

**Cyanobacteria.** Also known as blue-green algae. Found in soil and water and are able to fix nitrogen and photosynthesize.

**Cyclonic Storm.** Type of weather that results when cold air comes in contact with warm air in the mid-latitudes.

**Deciduous.** Refers to plants, usually trees and shrubs, that drop their leaves during non-growing seasons.

**Diurnal.** Change from day to night, such as temperature.

**Dwarf Shrub.** Small shrub with branches less than 12 in (30 cm) high.

**Ecotone.** A zone between two biomes or plant communities where type of climate and plants changes gradually.

**Ecotype.** A plant or animal that has not quite evolved to the species level but that has variations according to environment.

**Endemic.** Originating in and restricted to a particular geographic area.

**Energy Balance.** Comparison of the amount of energy received from the sun to the amount lost from Earth. Can be positive or negative. Also called radiation balance.

**Entisol.** Recently developed soils that have no differentiated horizons or layers.

**Epiderm.** Outer layer of leaf cells.

**Equable.** Refers to climate. Little variation between winter and summer temperatures.

**Ericaceous.** Refers to a member of the Ericaceae, a plant family consisting of heathers, bilberries, Labrador tea, rhododendrons, and so on.

**Evapotranspiration.** The combined processes of adding water vapor to the atmosphere through evaporation from the soil and bodies of water and from the passage of water out of plants through the stomata in their leaves.

**Evergreen.** Refers to plants, usually trees or shrubs, which maintain leaves all year. Leaves may be replaced throughout the year or in a flush during a single season, but the plant is never without live foliage.

**Fauna.** All the animal species in a given area.

**Flora.** All the plant species in a given area.

**Floristic Province.** A region with related plants that evolved in that area.

**Föhn.** A warm downslope wind.

**Foliose (lichen).** Leaflike in appearance. Compare with crustose and fruticose.

**Forb.** A broad-leaved, green-stemmed, nonwoody plant. One type of herb.

**Fruticose (lichen).** Shrubby, upright lichens. Compare with crustose and foliose.

**Genus (plural = genera).** A taxonomic unit composed of one or more closely related species.

**Giant Rosette.** Rosettes in tropical alpine environments that grow to large size.

**Gigantism.** Trait of some species to grow larger than normal.

**Glacial Drift.** Any rock and sediment carried and deposited by a glacier. See moraine, outwash, and boulder clay.

**Gleization.** The process of soil formation in bogs where undecayed plant material accumulates under conditions of poor drainage and cold temperatures. Forms a peaty surface underlain by clay.

**Graminoid.** Refers to narrow-leaved grass-like plants, including grasses, sedges, and rushes.

**Growing Season.** The length of time plants are able to grow, generally between the last frost of winter and first frost of fall, but may be related to seasonality of precipitation.

**Growthform.** The appearance or morphology of a plant that is adapted to particular environmental conditions. Examples include tree, shrub, and forb.

**Habitat.** The place where a species lives and the local environmental conditions of that place.

**Heath.** Small-leaved shrubs such as heathers and bilberries, which are members of the Ericaceae. Leaves are often drought adapted.

**Hemicryptophytes (Raunkiaer).** Plants that hold regenerating buds at the surface of the ground.

**Herb.** A nonwoody or soft, green-stemmed plant that dies down each year. May be annual or perennial. Broad-leaved herbs are called forbs. Grasses and sedges are called graminoids.

**Herbaceous.** Nonwoody plant that dies back every year.

**Hibernation.** When an animal's body temperature is reduced to that of the environment, resulting in a decrease in metabolism and need for energy. Used by animals to avoid a cold season.

**High Arctic.** Refers to arctic latitudes with high numbers, usually with a more severe climate than Low Arctic.

**Histosol.** Soils that are predominantly organic, usually formed by the gleization process and common in bogs.

**Humus.** Partially decayed plant and animal matter that occurs as a dark brown organic substance in soils. Important for conserving water and some plant nutrients.

**Inceptisol.** In humid climates, developing soil that has at least one distinctive horizon.

**Indicator Plant.** A plant typical of a biome that can be used to delimit the biome's extent in the absence of climate data.

**Inflorescence.** Flower stalk.

**Infrared radiation.** Energy that comes from the Earth. Also called longwave radiation or terrestrial radiation.

**Introduced (species).** A species transported accidentally or deliberately by humans beyond its natural distribution area. Also called alien or non-native.

**Inversion.** Air is warmer with increasing elevation, instead of the normal lapse rate of cooler temperatures.

**Koeppen.** A climate classification system.

**Krummholz.** Stunted trees at treeline, deformed by harsh climatic conditions.

**Lapse Rate.** Decrease in temperature as elevation increases. Averages 3.5° F per 1,000 ft (6.5° C per 1,000 m), but the rate is variable.

**Latitude.** Distance north or south of the equator, measured in degrees. The equator is 0° latitude. Low latitudes lie between 0° and 30° north and south; mid-latitudes lie between 30° and 60°, and high latitudes between 60° and 90°.

**Lichen.** A form of life composed of a fungus and an alga joined in a symbiotic relationship and classified as a single organism.

**Lifeform (Raunkiaer's).** A category of plant life based on morphology and the position of the renewal bud.

**Low Arctic.** Refers to arctic latitudes with lower numbers, usually with a less severe climate that High Arctic.

**Maritime.** The effect large bodies of water have on moderating seasonal temperature variations. Maritime climates do not have extremes of temperature. Compare with continentality.

**Mat Plant.** A low-growing, multistemmed plant that roots from stems that creep along the ground surface. Forms a dense cover.

**Mesophyll.** Interior leaf cells.

**Microclimate.** A small area with climatic conditions that are different from the general climate of the area.

**Microhabitat.** A tiny nook or habitat with specific environmental conditions that differ from that of the larger habitat in which it occurs.

**Moraine (glacial).** Unsorted accumulation of rocks and sediment deposited by a glacier, usually as hills. A type of till.

**Morphology.** Form and structure, size and shape of an organism. General appearance of an organism.

**Needle ice.** Ice that freezes and expands in soil, disrupting the surface.

**Nival.** Zone of permanent snow and ice on mountains. Snowline marks its lower limit.

**Nivation Hollow.** Depression in the ground surface where snow accumulates and the soil is moist.

**Nunatak.** A mountain peak that extends or extended above glacial ice, providing a place for plant or animal life during glaciation. May be refugia.

**Outwash (glacial).** Sediment that is sorted by size as it is deposited by water melting from a glacier.

**Paludification.** The process by which acidic, waterlogged conditions are produced and bogs are expanded because of the growth of mosses.

**Parent Material.** Sediment or rock from which soil is developed. Contributes to the mineral component of the soil.

**Perennial.** A plant that lives for several years and undergoes active growth each year.

**Periglacial.** Refers to cold areas subject to intense frost action and solifluction, often adjacent to glaciers.

**Permafrost.** A condition of permanently frozen ground.

**Phanerophyte (Raunkiaer).** A plant that holds its regenerating buds high, such as trees.

**Photoperiod.** The number of hours of daylight. Depending on latitude, it may change with the seasons.

**Photosynthesis.** The process by which green plants convert oxygen and carbon dioxide in the presence of sunlight to sugars. Energy in the form of visible light is transformed into chemical energy that can be used by living organisms.

**Physiology.** The metabolic functions and processes of organisms.

**Pleistocene.** Geologic time period when glaciers covered much of North America and Eurasia, from approximately 1.6 million to 10,000 years ago. Also called the ice age.

**Polar Desert.** The coldest and driest part of the Arctic or Antarctic tundra. Polar semi-desert is not quite as dry or as cold.

**Polyploidy.** Refers to more than the normal two sets of chromosomes.

**Refugia.** Areas that were ice free during the Pleistocene. Plants and animals could maintain populations in refugia.

**Regenerating bud(s).** Point(s) where new growth will form in the growing season.

**Rhizome.** A horizontal root structure that lies just below the surface.

**Rosette.** A growthform characterized by a basal whorl of leaves around a central stem or renewal bud. Can be flat to the ground or taller.

**Sclerophyllous.** Referring to leaves that are thick or waxy to prevent water loss.

**Scree.** Unstable slopes of small-size rocks, subject to downslope movement.

**Scrub.** A vegetation type characterized by sparse, small shrubs or stunted trees.

**Seasonality.** Difference in temperature or precipitation between winter and summer.

**Sexual Reproduction.** The formation of new individuals by the fusion of gametes (ova) and pollen in plants; egg and sperm in animals.

**Soil.** The uppermost layer on land. Composed of a mixture of mineral and organic materials in which plants grow.

**Soil Horizon.** A layer within the soil that is fairly distinct in terms of its chemistry, texture, and color.

**Solar Radiation.** Energy that comes from the sun, also called shortwave radiation or insolation.

**Solifluction.** Process by which water-logged soil slowly flows downslope.

**Species.** A group of sexually reproducing individuals that can produce viable offspring. The fundamental unit of classification in taxonomy.

**Spodosol.** Soil that usually forms beneath forest vegetation. Has distinctive horizons, especially a light-colored sandy surface with a darker, clay layer beneath.

**Stomata.** The leaf pores through which plants exchange gases with the atmosphere.

**Subalpine.** The zone slightly lower in elevation than alpine tundra, but with some characteristics of the alpine environment.

**Subarctic.** The latitude just south of the Arctic Circle, but with some characteristics of the Arctic. Subantarctic refers to just north of the Antarctic Circle in the Southern Hemisphere.

**Sublimation.** When ice or snow does not melt but skips the liquid state and goes directly into the gas form in the air.

**Substrate.** Surface material in which plants grow, including rock, soil, or sediments.

**Succulent.** A growthform that has specialized tissue in the stem, leaves, or an underground organ for the storage of water.

**Sun Angle.** How high the sun is in the sky during the day and a major factor in temperature. A higher sun angle imparts more warmth; a lower sun angle, less.

**Supercool.** Ability of a plant or animal to reduce its body temperature below freezing with no harm to the tissues.

**Talus.** Unstable steep slopes of large rocks, subject to downslope movement.

**Taproot.** A root that extends deep into the ground that provides access to water.

**Tarn.** Alpine lake occupying a cirque in an alpine glaciated environment.

**Taxonomy.** The science of describing, classifying, and naming organisms.

**Temperate.** Refers generally to the temperature patterns of the mid-latitudes, the temperate zone, where summers are warm to hot and winters are mild to cool. Not too cold or too hot.

**Tertiary.** A recent geologic time period, from approximately 1.6 million to 65 million years ago.

**Thallus.** The body of a nonvascular plant that is not differentiated into stem, root, or leaf.

**Tolerance Limits.** Refers to the extremes of environmental factors, such as cold, heat, drought, and snow depth, beyond which an individual species cannot survive.

**Torpor.** Reduction in an animal's body temperature for a short period of time, allowing the animal to avoid an unfavorable environmental period. The animal becomes torpid. It is not true hibernation because the body temperature is not reduced to that of the environment.

**Treeline.** The transition from forest to nonforested zones.

**Tropical Alpine.** Treeless zones in high mountains near the Equator or in the Tropics.

**Tropics.** The parts of Earth that lie between $23^1/_2°$ N and $23^1/_2°$ S or between the Tropic of Cancer and the Tropic of Capricorn.

**Tundra.** Refers to treeless zones in the Arctic, Antarctic, or alpine.

**Tussock.** A growthform of grasses and sedges in which individuals grow as tufts or clumps and form conspicuous hummocks on the ground.

**Ungulate.** Any hoofed mammal, such as a cow, elk, or ibex.

**Vascular plant.** Any plant with conducting vessels that move nutrients and water between roots and leaves. Includes flowering plants and ferns.

**Vegetation.** The general plant cover of an area, defined according to the appearance of the plants (grass, forest, and shrubs) rather than the actual species present. Compare with flora.

**Vegetative Reproduction.** The formation of new plants from pieces of the parent plant such as fragments of leaves, roots, stems, or rhizomes. Also called asexual reproduction. Includes cloning.

**Viviparous.** Refers to plants that produce plantlets or seed-like shoots instead of seeds.

**Zonation.** The occurrence of particular forms of life in distinct belts. These may be determined by latitude or elevation.

# Bibliography

## General

Constable, George. 1985. *Planet Earth, Grasslands and Tundra.* Alexandria, VA: Time-Life.

Ives, Jack D., and Roger G. Barry, eds. 1974. *Arctic and Alpine Environments.* London: Methuen.

Weatherbase. n.d. http://www.weatherbase.com.

Woodward, Susan L. 2003. "Tundra." In *Biomes of Earth,* 205–229. Westport, CT: Greenwood.

## Arctic and Antarctica Tundra

Bliss, Lawrence C. 1997. "Arctic Ecosystems of North America." In *Polar and Alpine Tundra,* ed. F. E. Wielgolaski, 551–683. Ecosystems of the World, 3. Amsterdam: Elsevier.

Bliss, Lawrence C. 2000. "Arctic Tundra and Polar Desert Biome." In *North American Terrestrial Vegetation,* eds. Michael G. Barbour and William Dwight Billings, 1–40. 2nd ed. Cambridge: Cambridge University Press.

Böcher, J., and P. M. Petersen. 1997. "Greenland." In *Polar and Alpine Tundra,* ed. F. E. Wielgolaski, 685–720. Ecosystems of the World, 3. Amsterdam: Elsevier.

Central Intelligence Agency. 1978. *Polar Regions Atlas.* Washington, DC: U.S. Government Printing Office.

Chernov, Y. I., and N. V. Mataveyeva. 1997. "Arctic Ecosystems in Russia." In *Polar and Alpine Tundra,* ed. F. E. Wielgolaski, 361–507. Ecosystems of the World, 3. Amsterdam: Elsevier.

Kanda, H., and V. Komárková. 1997. "Antarctic Terrestrial Ecosystems." In *Polar and Alpine Tundra,* ed. F. E. Wielgolaski, 721–761. Ecosystems of the World, 3. Amsterdam: Elsevier.

Wielgolaska, F. E. 1997. "Fennoscandian Tundra." In *Polar and Alpine Tundra,* ed. F. E. Wielgolaski, 27–83. Ecosystems of the World, 3. Amsterdam: Elsevier.

## Mid-Latitude Alpine Tundra

Billings, William Dwight. 2000. "Alpine Vegetation." In *North American Terrestrial Vegetation,* eds. Michael G. Barbour and William Dwight Billings, 537–572. 2nd ed. Cambridge: Cambridge University Press.

Campbell, J. S. 1997. "North American Alpine Ecosystems." In *Polar and Alpine Tundra,* ed. F. E. Wielgolaski, 211–261. Ecosystems of the World, 3. Amsterdam: Elsevier.

Grabherr, Georg. 1997. "The High-Mountain Ecosystems of the Alps." In *Polar and Alpine Tundra,* ed. F. E. Wielgolaski, 97–121. Ecosystems of the World, 3. Amsterdam: Elsevier.

Killick, D. J. B. 1997. "Alpine Tundra of Southern Africa." In *Polar and Alpine Tundra,* ed. F. E. Wielgolaski, 199–209. Ecosystems of the World, 3. Amsterdam: Elsevier.

Körner, Christian. 2003. *Alpine Plant Life of High Mountain Ecosystems.* 2nd ed. Berlin: Springer-Verlag.

Mark, A. F., and K. J. M. Dickinson. 1997. "New Zealand Alpine Ecosystems." In *Polar and Alpine Tundra,* ed. F. E. Wielgolaski, 311–345. Ecosystems of the World, 3. Amsterdam: Elsevier.

Miehe, Georg. 1997. "Alpine Vegetation Types in the Central Himalayas." In *Polar and Alpine Tundra,* ed. F. E. Wielgolaski, 161–183. Ecosystems of the World, 3. Amsterdam: Elsevier.

Zwinger, Ann H., and Beatrice E. Willard. 1972. *Land Above the Trees: A Guide to American Alpine Tundra.* New York: Harper & Row.

## Tropical Alpine Biome

Bussmann, Rainer W. 2006. "Vegetation Zonation and Nomenclature of African Mountains—An Overview." *Lyonia, A Journal of Ecology and Application* 11 (1): 41–66.

Coe, Malcolm James. 1967. *The Ecology of the Alpine Zone of Mount Kenya.* The Hague: Dr. W. Junk Publishers.

Diaz, A., J. E. Péfaur, and P. Durant. 1997. "Ecology of South American Páramos with Emphasis on the Fauna of the Venezuelan Páramos." In *Polar and Alpine Tundra,* ed. F. E. Wielgolaski, 263–310. Ecosystems of the World, 3. Amsterdam: Elsevier.

Hedberg, O. 1997. "High-Mountain Areas of Tropical Africa." In *Polar and Alpine Tundra,* ed. F. E. Wielgolaski, 185–197. Ecosystems of the World, 3. Amsterdam: Elsevier.

Locklin, Claudia. n.d. "Central Andean Dry Puna (NT1001)." Terrestrial Ecoregions. http://www.worldwildlife.org/wildworld/profiles/terrestrial/nt/nt1001_full.html.

Loope, Lloyd L. 2000. "Vegetation of the Hawaiian Islands." In *North American Terrestrial Vegetation,* eds. Michael G. Barbour and William Dwight Billings, 661–688. 2nd ed. Cambridge: Cambridge University Press.

Magin, Chris. n.d. "Ethiopian Montane Moorlands (AT1008)." Terrestrial Ecoregions. http://www.worldwildlife.org/wildworld/profiles/terrestrial/at/at1008_full.html.

Salcedo, Juan Carlos Riveros, and Claudia Locklin. n.d. "Central Andean Wet Puna (NT1003)." Terrestrial Ecoregions. http://www.worldwildlife.org/wildworld/profiles/terrestrial/nt/nt1003_full.html.

Patzeit, Erwin. 1996. *Flora del Ecuador.* Quito: Banco Central del Ecuador.

Vuilleumier, Francois, and Maximina Monasterio. 1986. *High Altitude Tropical Biogeography.* New York: Oxford.

Ziegler, Alan C. 2002. *Hawaiian Natural History, Ecology, and Evolution.* Honolulu: University of Hawaii Press.

# Index

The letter *f* following a page number denotes a figure. The letter *t* following a page number denotes a table.

**About the Author**

JOYCE A. QUINN is a professor in the Department of Geography at California State University–Fresno.